VALUE ANALYSIS
IN DESIGN
AND CONSTRUCTION

James J. O'Brien, P.E.

VALUE ANALYSIS IN DESIGN AND CONSTRUCTION

McGraw-Hill Book Company

New York   St. Louis   San Francisco   Auckland   Düsseldorf
Johannesburg   Kuala Lumpur   London   Mexico   Montreal   New Delhi
Panama   Paris   São Paulo   Singapore   Sydney   Tokyo   Toronto

Library of Congress Cataloging in Publication Data

O'Brien, James Jerome, date
Value analysis in design and construction.

Includes index.
1. Value analysis (Cost control)   2. Building—
United States.   3. Design, Industrial.   I. Title.
TS168.024   658.1′552   75-19169
ISBN 0-07-047566-0

1 2 3 4 5 6 7 8 9 0   KPKP   7 8 5 4 3 2 1 0 9 8 7 6

The editors for this book were Jeremy Robinson and Robert Braine,
the designer was Elliot Epstein, and the production supervisor
was George E. Oechsner. It was set in Palatino and Helvetica
by Progressive Typographers.

It was printed and bound by The Kingsport Press.

*DEDICATION to
WORLD OPEN UNIVERSITY
. . . and its new values in
professional graduate education.*

CONTENTS

PREFACE

This book is written for a team—a team which often does not work together. It is made up of the designers, constructors, value analysts, and owners of a construction project.

Value analysis has been an established technique for more than 20 years, and many in the construction field—designers and contractors—have been trained in the concepts and techniques of value analysis and value engineering. Only about half of those in the industry are aware of value analysis, perhaps less than 5 percent realize its potentials, and only about 1 percent are actively utilizing value-analysis techniques. This is an incredible situation. There are definite reasons for this apparent apathy, and the value analyst needs to understand and empathize with the balance of the team in order to turn them on rather than turn them off to value analysis.

The role of the design team is vital to the successful application of value analysis. It is during this phase that most consistent opportunities for major value-analysis implementation occur. Architects and engineers are constantly involved in value judgments during design, and this book is directed at an explanation of the compatibility of value-analysis techniques with the value judgments presently being made. Without this awareness, the design team can well view the imposition of value-analysis requirements as an affront to their professionalism.

Contractors are almost constantly aware of project situations in which poor value judgments were made. Even though most of the actual implementations of value analysis have been in the construction phase, the real potential lies in the ability of experienced contractors to participate in value analysis during the design phase. The increasing recognition of the potential of construction management is simultaneously opening the door to more value analysis. One of the problems in the construction field has been the dichotomy between design and construction.

Last, but most important, is participation by the owner. Value analysis, to a degree, can be implemented without additional cost on the part of the design team. However, to be most effective it needs the active support of the owner. To be willing to provide the support, the owner needs awareness of the responsibilities of the entire team—and the limitations of those responsibilities under the traditional fee structure.

Value analysis, effectively applied, should provide a means of communication between the members of the construction team. It is common sense applied creatively in a timely fashion.

The principal purpose of this book is to encourage meaningful value analysis at the project level by establishing a bridge between the discipline of value analysis and the roles of the key design and construction professionals as they now perceive them.

Value analysis specialists use the terms "value analysis" and "value engineering" interchangeably. However, the term engineering has a very definite preestablished meaning in the design and construction fields. Its application to value analysis would imply a narrow operation not touching many of the activities of the owner, the architect, and the contractor. For that reason, a conscious effort has been used to select the term value analysis as all-encompassing in this presentation. There are a few places where value engineering is inescapable—such as the VECP (value-engineering change proposal), which is a term well established in various federal organizations.

Because of the dichotomy in the two well-established terms value analysis and value engineering in their application to design and construction, there is another trend toward the use of a new term: the Public Building Service has renamed its Office of Value Engineering Office of Value Management.

James J. O'Brien, P.E.

ACKNOWLEDGMENTS

Larry Miles, CVS   for developing the discipline, providing leadership to SAVE, and for his generous advice on this book

Al Dell'Isola, CVS & Jim Hudson, CVS   for their trailbreaking in the construction field

Arthur Sampson (Administrator GSA) & Wally Meisen (Commissioner PBS)   for their continuing sponsorship of Value Management in public construction

Don Parker, CVS   for his trailblazing at PBS

Dale Daucher, CVS & Glenn Woodward, CVS   for their continuing Value Management program work at PBS

Maj. Gen. Bill Starnes, USA (ret.)   for sharing his experience in the Ohio River Division Corps of Engineers Value Engineering program

Paul Dobrow, CVS   for his continuing work in developing Value Engineering programs through the Office of the Chief of Engineers, U.S. Army Corps of Engineers

and

The Society of American Value Engineers (SAVE)

VALUE ANALYSIS
IN DESIGN
AND CONSTRUCTION

INTRODUCTION

Value analysis developed as a specific technique after World War II. The development work was done at the direction of the General Electric Company vice president of purchasing, Mr. Harry Erlicher, who observed that some of the substitute materials and designs utilized as a necessity because of wartime shortages offered superior performance at lower cost. He directed that an in-house effort be undertaken to improve product efficiency by intentionally developing substitute materials and methods to replace the function of more costly components. This task was assigned in 1947 to Lawrence Miles, staff engineer with General Electric Company. Miles researched the techniques and methodology available and utilized a number of proven approaches in combination with his own procedural approach to analysis for value. The value-analysis technique was accepted as a G.E. standard, and gradually other companies and governmental organizations adopted the new approach as a means of reducing costs.

In 1954, the Navy Bureau of Ships became the first Department of Defense organization to set up a formal value-analysis program, and Lawrence Miles was instrumental in the development of that program. The program was retitled "value engineering" (VE) to reflect the engineering emphasis of the Bureau of Ships.

In 1956, the Army Ordnance Corps initiated a value-engineering program, again with the assistance of G.E. personnel. This program has continued over the years, and the Army Management Engineering Training Agency offers value-engineering training as part of its curriculum.

In 1961, the Air Force became interested in the potential of value engineering as a result of the effectiveness of applications by Air Force weapons systems contractors such as G.E. This interest was one of the building blocks which, in 1962, led Secretary of Defense McNamara to place his prestige behind the use of value engineering, calling it a key element in the drive to reduce defense costs. (In that same year, Secretary McNamara gave great impetus to the development of network-based planning systems such as PERT and CPM by his mandate that these techniques be utilized in all project and program planning for the Department of Defense.)

To implement the directive of Secretary McNamara, value engineering was included as a mandatory requirement by the Armed Services Procurement Regulations (ASPR). The Armed Forces' definition of value engineering (Ref. 1) is: "A systematic effort directed at analyzing the functional requirement of the Department of Defense systems, equipment, facilities and supplies for the purpose of achieving essential functions at the lowest total cost, consistent with the needed performance, reliability, quality, and maintainability."

VALUE ANALYSIS IN CONSTRUCTION

Prior to the introduction of value engineering into ASPR, applications in the construction field were random and infrequent. Although the adoption of value engineering by the Department of Defense was oriented principally toward the purchase of materials, equipment, and systems, the change in ASPR automatically introduced value engineering to two of the largest construction agencies in the country: the U.S. Army Corps of Engineers and the U.S. Navy Bureau of Yards and Docks. In the first 10 years of use, the Corps of Engineers estimated savings of almost $200 million, with most of the savings the result of its own reevaluations of major projects. In that time period, 2,200 contractors had submitted cost-cutting value-engineering suggestions, of which 1,400 were accepted with the cash-shared savings of $7 million.

As the success of value engineering, principally in the construction phase, was documented, other federal agencies began tentative steps toward the adoption of value engineering. In 1965, the Bureau of Reclamation undertook value-engineering training for its engineering staff and in 1966 placed a value-engineering incentive clause in its construction contracts.

In 1967, the Post Office Department set up a value-engineering staff in its Bureau of Research and Engineering. In that same year, the Senate Committee on Public Works held hearings on the use of value engineering in the government at which many of the major agencies exchanged information on their utilization of value engineering. In 1969, the Office of Facilities of the National Aeronautics and Space

Administration (NASA) began formal value-engineering studies and training. The U.S. Department of Transportation published a value-engineering incentive clause to be used by its agencies. In that same year, the Public Building Service of the General Services Administration (PBS/GSA) set up its value-engineering staff.

Until 1972, the construction industry, in general, had only limited interest in value analysis or value engineering. In 1972, the twelfth annual conference of the Society of American Value Engineers (SAVE) emphasized the application of value analysis in the construction industry. Chartered in 1959, SAVE has had much to do with the evolution of value analysis, particularly in the federal establishment. Key SAVE members in the Washington chapter are predominantly from federal departments. The SAVE conference provided a forum where some 400 engineers, architects, and other industry members heard the specifics of progress which had been made (Ref. 2). Jurisdictions discussing value engineering included the Corps of Engineers, the GSA, the Veterans Administration, NASA, the Department of Health, Education, and Welfare (HEW), the Postal Service, and the Department of Defense, as well as the states of Pennsylvania, New York, and Massachusetts, and the cities of Chicago, Savannah, Portland, and Washington, D.C. With the exception of the Corps of Engineers, most of the action was in the areas of training, seminars, and staff development.

At the thirteenth conference, which was held in Chicago, it was estimated that more than half of the attendees were architects, engineers, and contractors. Maj. Gen. W. L. Starnes, division engineer of the Ohio River Division, described the activities of the Corps of Engineers in his area (see Case Study N). For the first time, a major application by a contractor was discussed. Edward Poth, vice president of Paschen Contractors, described a $4 million value-engineering saving which his firm in cooperation with the engineer, owner, and architect had achieved on two $55 million projects (Ref. 3).

The major federal agencies involved in the application of value engineering in facilities programs have cooperated in the exchange of information and the development of different approaches. Because of the different charters authorizing activities of the organizations, their application of value analysis varies. The Department of Health, Education, and Welfare, for instance, has direct control of several hundred million dollars in construction value per year but indirect control of perhaps more than a billion dollars in an average year through its grant and support programs for hospitals and education. HEW has established a cost-avoidance program (Ref. 4) which they indicate has saved $333 million in 3 years. Of this saving, 30 percent was in the federal share of supported programs, with the balance distributed between state and local agencies. The significance of these savings, however, is not limited to their dollar value. The exposure of architects, engineers, and contractors to actual value-engineering savings during construction reinforces the concept—and provides a positive momentum to the approach. Where the Department of Defense was responsible in 1962 for the first major introduction of value engineering to the construction industry, in 1972 the GSA under Administrator Arthur Sampson clearly took leadership of a positive program to expand dramatically the utilization of value analysis in both design and construction. Under Administrator Sampson, GSA has actively undertaken the evaluation and implementation of several improved management techniques to be applied to the implementation of design and construction for federal nonmilitary building facilities. As the purchaser of hundreds of millions of dollars in annual construction value, these efforts toward improved management have demanded the attention of the entire construction industry. One of these techniques is value management.

On March 1, 1972, the GSA required that most new architectural, engineering, and construction management contracts contain value-engineering clauses (Ref. 5).

Engineering News-Record noted that efforts to mobilize a governmentwide value-engineering program are off to a slow start (Ref. 6). An organizational meeting was called by GSA Administrator Sampson for mid-April 1974, with an invitation to heads of about 20 agencies to discuss the establishment of a federal value-management council to develop and coordinate value-engineering efforts for all agencies. However, a more organized approach can be anticipated, since the General Accounting Office (GAO) has now been reported as urging a more coordinated use of value analysis and chiding agencies that appeared to be lagging. *Engineering News-Record* indicated that GAO stopped short of suggesting legislation to require value-engineering clauses (Ref. 7), but the report did highlight the advantages of the approach. Further, GAO suggested that agencies should aggressively publicize actions which result in savings, both within and outside the agencies.

A more complete review of the GAO report entitled "Need for Increased Use of Value Engineering, a Proven Costsaving Technique in Federal Construction" (Ref. 8) indicated that the report was based on an evaluation of value-engineering activities in 10 federal agencies. Four agencies have had value-engineering incentive programs for about 10 years. Two recently initiated programs and four (NASA, AEC, HEW, and TVA) have no incentive programs. The report indicated that the GSA won highest praise for its value-engineering incentive clause which is relatively simple, easy to understand, and directed specifically to construction. These features were contrasted with the incentive clause used by the Department of Defense and included in its ASPR, which runs to some pages. GAO considered this too complicated and generalized for use in construction.

The GAO noted "there is a need to circulate proven value engineering proposals, both within and among federal agencies. A major benefit of the value engineering proposal is its potential for repetitive use on other projects" (Ref. 8).

LEGISLATION

Presently, there is no legislation which requires the use of value engineering/value analysis on federal projects. Nevertheless, the individual agencies have made substantial progress in that direction through executive regulations. The legislative sector has been interested for some time, however. At the twelfth annual SAVE meeting, Senator Jennings Randolph of the Committee on Public Works indicated that hearings would be held on the use of value engineering to reduce construction costs. With new construction expenditures estimated at $32 billion in public works, cost savings in the range of 5 percent through value engineering would save at least $1 billion on these projects.

Two of the leading agencies reported on their progress at the fourteenth annual SAVE meeting (Ref. 9). Paul V. Dobrow, chief of the Corps of Engineers value-engineering office, estimated that the total cumulative savings through value engineering in the Corps were now almost $234 million. Commissioner Larry F. Roush of PBS/GSA indicated that GSA had netted savings of $4.53 for every dollar spent administering its value-engineering program, for a total savings to GSA of $1.8 million in fiscal 1973.

A prime supporter of value engineering, Representative Larry Winn, Jr., of Kansas, a former building contractor, indicated that he intends to introduce legislation which would require value engineering (or as he prefers to term it, "value management") of all government construction contracts.

V-E OPPORTUNITIES

The Department of Defense conducted a study in 1965 to determine the predominant sources of opportunity for value engineering (Ref.

10). From a sample of 415 successful value-engineering changes, the study identified seven factors which were responsible for about 95 percent of the savings actions. While these factors applied principally to industrial products, shipbuilding, and weapons systems, all seven reasons apply to the construction industry as well. These seven predominant reasons or opportunities for application of value-engineering were:

1. Recognition of advances in technology
2. A questioning of specifications for value
3. The application of additional design skills, ideas, and information available but not previously utilized
4. Recognition of a change in user's needs
5. A design tuned to value
6. Feedback from field test or use
7. Deficiencies in design

SUMMARY

The potential for application of value analysis in the construction industry is tremendous. The way has been pointed out by the technique's developers, who, working through their Society of American Value Engineers, continue to conduct seminars and conferences to educate others about its potential and actively work on the implementation of value analysis, which may soon be legally required in public contracts supported by federal funds. These pioneers have broken the trail and shown the way. Much remains to be done before it can be said that the construction industry has recognized *and* implemented a substantial portion of the opportunities offered by value analysis.

REFERENCES

1. "Principles and Applications of Value Engineering," Department of Defense Joint Course, vol. 1, U.S. Government Printing Office, Washington, D.C.
2. *Engineering News-Record*, June 22, 1972, p. 180.
3. *Building Design & Construction*, July 1973.
4. *Civil Engineering*, Nov. 1973.

5. *Engineering News-Record,* Feb. 22, 1973.

6. *Engineering News-Record,* May 9, 1974.

7. *Engineering News-Record,* May 23, 1974.

8. *Building Design & Construction,* July 1974.

9. *Engineering News-Record,* May 9, 1974.

10. Public Building Service/GSA Handbook P8000.0, *Value Engineering,* U.S. Government Printing Office, Washington, D.C.

FUNCTIONAL ANALYSIS

Lawrence Miles describes value analysis as "a disciplined action system, attuned to one specific need: accomplishing the functions that the customer needs and wants. . . ." (Ref. 1). He further indicates that "the basic function is straightforward to determine, and any work done before that has been accomplished is wasted effort. . . ."

The Public Building Service describes functional analysis in this way: "A user purchases an item or service because it will provide certain functions at a cost he is willing to pay. If something does not do what it is intended to do, it is of no use to the user and no amount of cost reduction will improve its value. Actions that sacrifice needed utility of an item actually reduce the value to the user. On the other hand, expenditures to increase the functional capacity of an item beyond that which is needed are also of little value to the user . . ." (Ref. 2). The Value Management Office of PBS sees anything less than the necessary functional capability as unacceptable—and anything more as unnecessary and wasteful. A project or part of a project which is to receive functional analysis is addressed with the six basic questions of value analysis:

What is it?

What does it do?

What is its worth?

What does it cost?

What else would work?

What does that cost?

WHAT IS IT?

A prerequisite to functional analysis is the selection of the function to be analyzed. Figure 2-1 shows a work-breakdown analysis for a weapons system. Selection of the optimum function for analysis depends upon the stage of development of the project as well as the relative value of the different components.

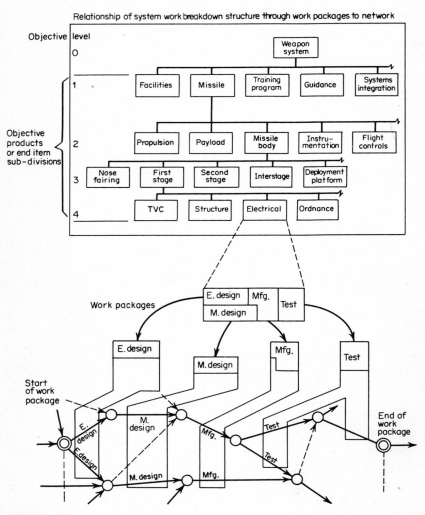

FIGURE 2-1
Work structure network. (U.S. Navy Special Projects Office, Government Printing Office.)

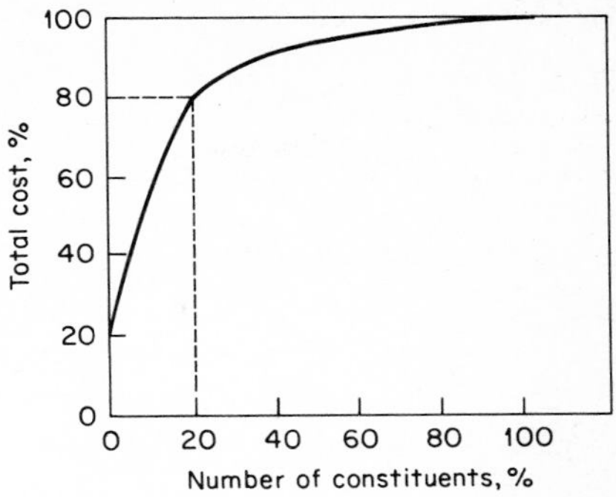

FIGURE 2-2
Pareto's law of distribution. (Public Building Service GSA Manual P 8000.1, *Value Engineering*.)

At a very early stage, the weapons system could be evaluated on a functional-analysis basis as an overall concept. However, if the development of the system is well advanced, the opposite end of the spectrum would be available for analysis, such as the different work packages which make up the electrical portion of the first stage of the missile body.

Within the area to be analyzed, selection of the functions which should be considered can be guided by the Pareto distribution shown in Fig. 2-2. Pareto's law of distribution, often referred to by value analysts, suggests that 20 percent of the items in any complex thing account for 80 percent of the cost. As will be noted on the section on cost, this distribution does not exactly hold for construction projects. However, it is axiomatic that a minority of the components of a project make up the majority of the cost. Within this minority of high-cost areas, the best opportunities can be found for value analysis. The identification of the high-value items narrows the number of viable opportunities for value analysis which remain in a project. Analysis of the high-value areas produces the "best bang for the buck." Standard cost-breakdown models for types of projects can be utilized in identifying the potential areas for best results.

It is the consensus of value specialists that the most productive areas of opportunity should be selected for analysis—and these areas are inherently the most expensive ones.

Knowledgeability regarding the cost makeup of components of a project is important. Without this, an uninformed analyst might assume that "mechanical work" at 25 percent of the project cost is a more opportune area for functional analysis than "superstructure" at 16 percent. However, mechanical work breaks down into subpackages which would require multiple functional analysis—while the superstructure may be analyzed in one single study.

WHAT DOES IT DO?

This is the key value-analysis question. It requires definition of the function of the item under study. The method of this definition is prescribed by established value-analysis procedure as two words, a *noun* and a *verb*.

The simplicity of the approach is deceptively simple. Selection of the proper two-word description is often quite difficult, requiring comprehensive understanding of the item under study.

The functional description is not necessarily correct when it is the most obvious—and therein lies the potential for successful analysis. The analyst is not limited to a single two-word description. In fact, a series of descriptions can apply to the same item. However, *only one* of these descriptions is the basic function. The other descriptions, necessarily, become secondary functions. Secondary functions may be important, but they are not controlling.

It is quite usual to find that many have assumed that the basic function is really one which had always been considered a secondary one.

Verbs Some typical verbs which can be used to describe construction functions are:

absorb	enclose	protect
amplify	filter	reduce
apply	generate	reflect
change	hold	reject
collect	improve	separate
control	increase	shield
conduct	insulate	support
create	interrupt	transmit
decrease	prevent	

Nouns measurable:

circuit	force	power
contamination	friction	pressure
current	heat	protection
damage	insulation	radiation
density	light	repair
energy	liquid	voltage

flow	noise	water
fluid	oxidation	weight

Nouns aesthetic:

appearance	effect	prestige
beauty	features	style
convenience	form	symmetry

The difficulty of identifying in a two-word phrase is often substantial. Accordingly, nouns such as thing, part, article, device, or component generally indicate a failure to conclusively define. The noun should be measurable or capable of quantification.

For instance, a water service line to a building could have the function "provide service," but the term "service" is not readily measurable. A better definition would be "transport water." The noun in this definition is measurable, and alternatives can be determined in terms of the quantity of water to be transported, or in special situations such as laboratories, in terms of the quality of water being transported.

The functional definition must concern itself with the type of use as well as the identification of the item. A piece of wire might "conduct current," "fasten part," or "transfer force," depending upon the specific utilization.

The spartan restraint involved in holding the description to two words can be difficult to apply. There is often a temptation to utilize a slightly broader definition by insertion of additional adjectives to condition the noun. Some value analysts permit this variation, but the tried and tested approach insists upon the two-word abridgement. The rigorous application of the two-word functional definition often discloses a factual view of the project category which was subconsciously available but not consciously realized.

For instance, one perception would be that it is not the door itself but the doorway or the accessway which is the true basic purpose for the specification of a door at a given location. The overall context of the application has to be considered. In an office building, the basic purpose of a doorway is to "provide access," while in a prison, this might be to "control access." An interior door at a fire wall does "provide access," but its basic purpose is to "control fire." A doorway at a classroom entrance might be there to "control noise."

At a detail level, doors throughout a building would be separate categories. To evaluate the generic category, the subbreakdown would be an important factor. Figure 2-3 describes some of the functional

VALUE ENGINEERING FUNCTIONAL ANALYSIS WORKSHEET

PROJECT _Courthouse_

ITEM _Doors_

BASIC FUNCTION _Control Access_

NAME _____

TEL. NO. _____

DATE _____

QUANTITY	UNIT	ELEMENT DESCRIPTION	FUNCTION VERB	NOUN	KIND	EXPLANATION	WORTH	COST
		Exterior doors	Provide	Security	B	Control of access mandatory		
			Exclude	Weather	S	Alternates possible		
			Express	Prestige	S	Image to the public		
			Provide	Visibility	S	Entry doors		
		Office doors	Exclude	Noise	B	Judge's chambers		
			Contain	Noise	B	Typing pool		
			Enclose	Space	B	Public areas		
		Area doors	Provide	Visibility	S	" "		
			Exclude	Fire	B	Exit passageways		
			Contain	Fire	B	Mechanical rooms		
			Control	Air	B	HVAC zone perimeter		
			Control	Traffic	B	Key zone points		
		Security doors	Control	Prisoners	B	Secure passage to courtrooms		
			Contain	Prisoners	B	Cell doors; cell block		
			Exclude	Public	B	Doors from public areas		
			Exclude	People	B	Elevator Shaft doors		
		Other	Provide	Privacy	B	Bathroom doors		

FIGURE 2-3
Functional definitions for doorways in a complex courthouse structure.

definitions for doorways in a complex courthouse structure. The cost of certain doors, such as the elevator shaft doors, would be included in other systems. However, a subset of the system of doors is the door bucks in which they mount. That system would be included in the hollow metal or miscellaneous metal subcontract.

The item under study can, and should, have different levels of indenture as identified in the work-breakdown structure. Figure 2-4 illustrates a breakdown structure for a fire alarm system prepared by PBS. The second level of indenture breaks down the system into two sectors: person and equipment. This identifies the system as semiautomatic. The table in Fig. 2-4 shows the functional definition developed with these various levels of indenture. At the third, or detailed, level the basic function of the bell was selected as "make noise," which would permit a greater latitude in the development of alternative ways of effectively making noise not limited to the bell.

WHAT IS IT WORTH?

Worth is a measure of value and represents the least expenditure required to provide the function required as defined by the functional definition. The evaluation may be specific but is quite often subjective. A method of deriving worth is identification of the cost of comparable items which would provide a similar service.

Many value analysts use the seven classes of value defined by Aristotle: economic, moral, aesthetic, social, political, religious, and judicial. The definition used by Mudge (Ref. 3) is: "Value as used in the systematic approach is defined as: the lowest cost to reliably provide the functions of service at the desired time and place and with the essential quality." He goes on to provide a number of subcategories: use value, esteem value, cost value, and exchange value. Within these categories, only use value is capable of a quantitative evaluation. However, in a facilities project, esteem value would be equivalent to aesthetic value and is a very important consideration.

If a building or facility is to be evaluated only on its use value, the worth described would be that of an engineered shelter. However, the human interreaction to and with a building is a subtle but very signif-

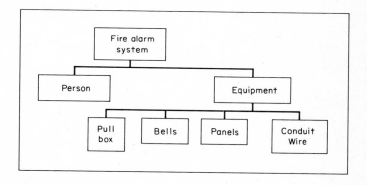

Level of Indention	Component	Functions	Classification B = Basic S = Secondary
1	Fire alarm system	Make noise Detect fire Protect building	B B S
2	Person	Detect fire Pull lever	B S
	Equipment	Make noise Transfer signals	B S
3	Pull boxes	Break circuit	S
	Bells	Make noise	B
	Panels	Provide power Control circuits	S S
	Conduit & wire	Transmit signal Transmit power	S S

FIGURE 2-4

Breakdown structure for a fire alarm system. (Public Building Service GSA Manual P 8000.1, *Value Engineering*.)

icant consideration. It is a relationship which may, in many ways, determine the volume of utilization of the facility. In terms of life cycle, the amount of utilization has a high impact upon the true worth. For instance: Two similar buildings are compared and have similar purposes and functions; one building has been constructed at the lowest first cost with no intentional aesthetic value added, while its companion has had aesthetic treatment designed to attract more users. If it is assumed that perhaps twice as many people will utilize the more attractive facility over its lifetime, and the additional cost of the aesthetics is less than the initial cost of the engineered shelter, a true quantitative value could be assigned to the cost of the aesthetics.

Aesthetic value might also be termed attractiveness value, wherein attraction is considered a real dollar-and-cents return. A supermart which attracts no customers is a very expensive investment, regardless of how low the initial cost has been.

The value of aesthetics cannot be dismissed or disregarded. In one value analysis of a major project (Ref. 4), the construction manager found the aesthetic value difficult to incorporate. After one value-analysis seminar, the construction manager was quoted as saying that suggestions made as part of the value analysis could save the GSA $1.4 million on an $11.5 million project, or about 12 percent. He then went on to say that "although limestone will cost $250,000 more, it would be aesthetically more pleasing than precast concrete . . . and we would still be saving over $1 million." The construction manager appeared to recognize the value of attractiveness but relegated it to a secondary function.

Proper application of value analysis should add to the willingness of well-informed owners to invest in attractiveness as a function.

WHAT DOES IT COST?

When the area has been defined enough to identify quantities, these can be costed and compared with worth. The result is a value index found as follows:

$$\text{Value index} = \frac{\text{worth}}{\text{cost}} = \frac{\text{utility}}{\text{cost}}$$

Accordingly, value may be increased by improving utility with no change in cost, retaining the same utility for less cost or combining an improved utility with a decrease in cost.

The choice of worth may be reconsidered when the cost of the area under study is established. In discussing something as basic as the worth of a bolt (Ref. 5), Heller suggests the results may be affected when the analysts are advised that this particular bolt holds on the wings of a transport plane. Engineers who would assign a certain dollar value to the worth of a bolt tend to increase the worth when they are asked to consider that they are passengers on an airplane in which that particular bolt holds on the wings. However, their original evaluation included a requirement that the item meet functional requirements. Implicit in this requirement is the understanding that the bolt is not allowed to fail on the service stated. Accordingly, the worth value for this bolt should not include an emotional value of the consequences of possible failure of the bolt.

The worth figure should be kept as basic as possible so that the value index theoretically cannot be above 1. In fact, if a cost figure of less than the worth can be achieved, and is a viable figure, the initial worth figure was incorrect. The value index provides a reasonable indication of the premium which an alternative approach is costing.

Some analysts attempt to hold the worth at the very lowest possible minimum—at a point which is below that which would be acceptable or feasible. One example in evaluating a pencil equates its worth to a nail which "makes marks." If the pencil is a drafting pencil, this can hardly be a realistic comparison. A more reasonable comparison would be the lowest acceptable quality of drafting pencil at the proper weight purchased in suitable quantities compared with, for instance, a lead holder. The cost of the lead holder, which is refillable, would be compared on a life-cycle basis with the cost of the number of equivalent pencils to determine a value index.

A common example used for worth in value analysis is the worth of a tie clip. If the function is really "hold tie," the value is perhaps equivalent to that of a paper clip. However, tie pins and tie clips usually cost much more than a paper clip. The worth is really in the appearance of the tie pin or tie clip, having a prestige value to the purchaser. For those users, however, who are really interested in "hold tie," a cloth loop sewn to the inner face can be used to hold the two sections of tie together when it is tied, at a cost of not much more than a paper clip. The function must be considered in terms of whether the user is really interested in prestige as a basic function or in "hold tie" as the basic function.

For construction projects, experience factors available from sources such as the McGraw-Hill Information Services (F. W. Dodge) which include case histories of equivalent projects can be a guideline. In comparing worth and cost, comparison figures should be adjusted to the same time reference to adjust for inflation and escalation factors.

Cost-worth comparisons must include all necessary items. For instance, in Ref. 6 the life safety systems for two Hyatt Regency Hotels are described. These systems are directly related to the trademark of the Hyatt Regency Hotels—a lobby rising to the roof with a revolving restaurant located on the top of a structural elevator core rising through the lobby. The sophisticated life safety systems required as a result of these innovations substantially add to the cost of the hotel, generally in the neighborhood of 4 percent of the total construction cost. Assignment of a worth to these hotels which did not include the cost of these trademarks would be unrealistic, since the trademarks are an absolute requirement of the owner.

WHAT ELSE WOULD WORK?

After the establishment of the functional definition and assessment of its worth and cost, the next key stage is the application of creativity to determine alternatives which would also perform the same function. Creativity is one of the essential techniques of value analysis. It is one of those existing attitudes important to value analysis which value analysts have incorporated into the total value-management approach.

It is human nature for people to resist change—to resist either stopping things which they are doing or initiating new ideas. The development of alternative approaches to meet functional analysis requires a breakdown of this inertial resistance to change.

There are many roadblocks to the creative approach, including but certainly not limited to the following:

1. Fear of making a mistake or appearing to make a mistake

2. Unwillingness to change the accepted norm

3. A desire to conform or adapt to standard patterns

4. Overinvolvement with the standard conceptions of function

5. Unwillingness to consider new approaches

6. Unwillingness to be considered rash or unconservative

7. Unwillingness to appear to criticize, even constructively

8. Lack of confidence resulting from lack of knowledge

9. Overconfidence because of experience, however limited

10. Unwillingness to reject a solution which has previously been shown to be workable

11. Fear of authority and/or distrust of associates

12. Unwillingness to be different

13. Desire for security

14. Difficulty in isolating the true functional requirements

15. Inability to distinguish between cause and effect

16. Inability to collect complete information

The roadblocks often manifest themselves in the form of subtle objections or standard platitudes, such as:

1. It may be okay, but management will never buy it.

2. We tried that approach before, and it didn't work.

3. Somebody tried that approach before, and it didn't work.

4. It might work, but it's not our responsibility.

5. It might work for someone else, but our requirements are unique.

6. It sounds okay, but we're not ready to progress that rapidly.

7. It sounds good, but we don't have enough time to implement the new approach.

8. Let's assign it to a committee.

9. Management might agree, but the union would be against it.

10. It's against company policy (we think).

11. We don't have the authority to make this change.

12. It's a good idea, but it would never work here.

13. It's never been tried before.

A key technique in the development of creative alternatives is the use of a think-tank or brainstorming approach. The group is used to develop ideas without evaluating them. The creative session is not limited to ideas which will probably work, but to a free-form type of idea listing. The idea is to break through typical inhibitions and restraints and list even outlandish ideas. Rube Goldberg would feel right at home in this kind of creative session.

In brainstorming, the starting point is the two-word abridged function definition. Ideas which are abstract, even humorous, are encouraged, because through stream-of-consciousness thinking and free associating, these ideas may lead to more practical considerations.

The creative session has its own set of ground rules to encourage free and open thinking. Not only is evaluation and judgment deferred, but also critical response is discouraged until the entire listing has been developed.

The functional A. is a keystone of V.A. but must be preceeded by the collection of information & followed by the creative process

SUMMARY

The basic value questions posed at the beginning of this section are asked many times in the course of a value analysis. The functional analysis is the keystone of value analysis, but it must be preceded by the collection of information (including cost) and followed by the creative process. The next section describes the stages of value analysis known as the job plan.

REFERENCES

1. Miles, L. D., *Techniques of Value Analysis and Engineering,* 2d ed., McGraw-Hill, New York, 1972, p. xvi.

2. Public Building Service/GSA Handbook P 8000.1, pp. 3-1 to 3-13.

3. Mudge, A. E., *Value Engineering,* 1971, McGraw-Hill, New York, 1971, p. 62.

4. *Building Design & Construction,* June 1974.

5. Heller, E. D., *Value Management: Value Engineering and Cost Reduction,* Addison-Wesley, Reading, Mass., 1971, p. 27.

6. *Engineering News-Record,* Sept. 12, 1974, p. 34.

FIGURE A-1
Functional analysis of the seven basic fire safety subsystems. ▶

CASE STUDY A
FUNCTIONAL ANALYSIS
AND CREATIVITY SESSION

The Public Building Service undertook the design and construction of three major payment centers for the Social Security Administration (SSA). These centers are located in Philadelphia, Chicago, and San Francisco. The combined value is in excess of $100 million.

The design of the facilities was under the direction of two executive architects, and a design team was located at each city. The entire program was under one construction manager, the Turner Construction Company. The author, Fred C. Kreitzberg, and Glenn Woodward (then a consultant, now value management program coordinator for PBS) conducted a 40-hour value-engineering seminar for a team made up of 20 representatives from the architect/engineer group, 9 representatives from PBS and SSA, and 6 members of the construction-management staff, including the 3 noted.

The group, after basic instruction, broke into five teams. This case study describes the work of the team investigating fire safety.

Figure A-1 describes the seven subsystem approaches to the basic purpose, which was the protection of life.

Figure A-2 shows a further analysis of the more detailed level of the seven subsystems.

VALUE ENGINEERING FUNCTIONAL ANALYSIS WORKSHEET

PROJECT *SSA Fire safety systems* NAME _____

ITEM _____ TEL. NO. _____

BASIC FUNCTION *S: Protect Life (B)* DATE _____
Protect Property (S)

QUANTITY	UNIT	ELEMENT DESCRIPTION	FUNCTION VERB	NOUN	KIND	EXPLANATION	WORTH	COST
		Basic purpose	*Counteract*	*Fire*	*B*			
		Safety subsystems						
		1.	*Reduce*	*Inflammables*	*B*			
		2.	*Detect*	*Fire*	*B*			
		3.	*Advise*	*People*	*B*			
		4.	*Protect*	*People*	*B*			
		5.	*Remove*	*People*	*B*			
		6.	*Maintain*	*Services*	*S*			
		7.	*Fight*	*Fire*	*B*			

PROJECT SSA Fire Safety NAME _____

ITEM _____ TEL. NO. _____

BASIC FUNCTION 1. Reduce Inflammables; 2. Detect Fire; DATE _____
3. Advise People

QUANTITY	UNIT	ELEMENT DESCRIPTION	FUNCTION VERB	NOUN	KIND	EXPLANATION	WORTH	COST
			Limit	Combustibles	B	Design to this function		
			Evaluate	Furnishings	S			
			Eliminate	Fuel	B			
		1. Reduce Inflammables	Microfilm	Files	S	Reduces bulk		
			Eliminate	Equipment	S	Reduces fuel reqt.		
			Fireproof	Files	S			
			Prohibit	Smoking	S			
			Operate	System	S			
		2. Detect Fire	Provide	Detectors	S			
			Provide	System	S			
			Alert	Firefighters	S			
			Provide	Alarm	B	Should tie to Detection		
		3. Advise People	Provide	Instruction	S			
			Conduct	Firedrills	S			

(a)

PROJECT SSA Fire Safety NAME _____

ITEM _____ TEL. NO. _____

BASIC FUNCTION 4. Protect People; 5. Remove People DATE _____

QUANTITY	UNIT	ELEMENT DESCRIPTION	FUNCTION VERB	NOUN	KIND	EXPLANATION	WORTH	COST
			Provide	Refuges	B			
			Sealoff	Fire	B			
		4. Protect People	Sealoff	Smoke	B			
			Provide	Air	B			
			Fire Proof	Structure	B			
			Program	Elevators	B			
		5. Remove People	Provide	Stair-Towers	B			
			Provide	Routes	B			
			Provide	Backup	B			
		6. Maintain Services	Sealoff	Fire	B			
			Zone	Controls	B			
			Provide	Equipment	B			
		7. Fight Fire	Provide	Firemen	B			
			Provide	Extinguishing	B			

(b)

FIGURE A-2
Analysis of a detailed level of the basic subsystems.

Table A-1 Value-Engineering Brainstorming List

PROTECT LIFE - PROTECT PROPERTY

Remove People
Fire Proof Everything
Remove Property
Shield People
Shield Property
Wear Protective Clothing
Sprinkler
Remove Smoke
Stairways For Each Floor
Exit Chutes /Slides
Blanket Fire
Large Capacity Elevators
Fire Poles
Fire Drills
Breathing Apparatus
Fire Extinguishers
PA System
No Openings In Bldg. Surface
Fire Warden /Patrol
Smoke Detectors
Heat Detectors
Restrict Flam. Furnishings
Exits For Fire Only
Exhaust Smoke
Neutralize Smoke
Dissipate Heat
Instructions
Fire Hoses
Garden Hoses
Outdoor Elev. For Firemen
Central Control
CO_2
Halon
Private Fire Dept.
Outside Fire Tower
Emergency Power
Helicopter
Closed Circuit T.V.
Contain Smoke
Refrigerate Bldg.
Compartmentation
Prohibit Smoking
Eliminate Fuel
Eliminate People
Water /Sand } Bucket
No Wiring
Water Blanket /Pitch Floor
No Gas

Floor Seal-Off; H_2O Tight Hatches
Flexible Fire Shield - To Isolate Bay
Flame Extinguishment Built into Carpet
H_2O Walls To Contain Fire
Floor Drains - Recirculation
Central CO_2 Plant
Mechanical Back-Up
Smoke Bldg
Back Burning
Fire Breaks
Single Floor Bldg
Site Near Fire Dept.
Site Near Natural H_2O
Precipitate Cloudburst
Integrate Fire Sprinklers to Total Energy System
Sprinklers w/ Independent H_2O Source
Run H_2O In Electrical Conduit
H_2O In Cells In Structural Floor
Fountain Columns
Sprinklers In Floor
Tubular Columns Filled With H_2O
Dry Ice Blanket Sys.
Vertical Curtains
Pressure Differential
Smoke Proof Towers
Collapsable Stairs In Elev. Shafts
Pressurize Elev. Lobbies
Over Design Sprinklers For Flexibility

Circular Slides
Bridges To Neighbor Bldg.
Increase Population To Decrease Detection Time
Individual or Desk Extinguishers
Dispose of Burning Material
Heat /Smoke Sensitive Windows /Open
Break Out Panels
Ladders Between Floors

Floor Hatches
Refuge Areas
Program Elevators
Fire Doors To Elev. Shafts
Piggy Back Elev. For Fires
Interior Fire Walls
Discontinue Elev. Service to Fire Floors
Rocket Pack
Auto. Closing Files on Heat Sensitive Basis
Parachutes
High-Lines To Next Bldg.
Fire Engine with Evac. Capability
Fire Locker For Firemen
Microfilm Files
Blow Out Fire
Use Plumbing Fixtures To Fight Fire
Remote Mechanical Equip.
Restrict H_2O Flow to Fire
Roof H_2O Tank
Flood Roof or Top Floor From Cooling Tower - Lump To The Exting. Sys. /Floor
Fire Proof File Folders
Total Module Construction
Train People To Fight Fire
Auto. Window Washer - Fire Fighting
Cooling Medium In Structure
Increase Relative Humidity
Explosion Vents
Heat Treated Glass for Exit & Entrance
Exterior Ladder - Movable
Silo
Ropes From Windows
Survival Areas
Moat /Foam Pad
Foam Sys.
Pool on Roof
Smoke Mask
No Building
Cold Air Into Stacks
Smoke Proof Stairs
Stack For Smoke Evac.
Integrate Sprinkler & Lights
Collect Rain H_2O
Sprinklers + Window Shades To Control Solar Heat

VALUE ENGINEERING PROPOSAL SUMMARY		DATE
		2/18/72

TO:	FROM:	PHONE NO
G.S.A.	Team #5	

ITEM NAME

Fire Safety Systems

COMPONENT OF	QUANTITY	STUDY SPAN:
SSA payment center		START 2/14/72
		COMPLETE 2/18/72

FUNCTION OF ITEM:	ESTIMATED SAVINGS:
VERB NOUN	
Protect Life & Property	

| DESCRIPTION OF PRESENT DESIGN

There is no "present design" since the buildings are presently under concept design, and the fire safety system for these buildings is undetermined. SSA's "Statement of Requirements" basically requires the building to provide fire safety such that "Any ignition will not affect SSA operations or personnel."

Because of the limited time available for value engineering, the team restricted the analysis to general office occupancy above the ground floor. Similar analysis should be made for the special purpose areas such as: files, computer room, cafeteria, auditorium, and mechanical rooms.

The Fire Safety Criteria for general office area was established as follows:

     1.   No loss of life
     2.   Fire severity of two hours
     3.   Assure no greater than 25% propagation to
          more than one work station.

The ultimate function of a fire safety system was designated as "Protection of Life and Property." This was broken down into three basic functions as follows:

     1.   Provide Egress
     2.   Control Combustion
     3.   Prevent Combustion

These basic functions were further broken down into secondary functions as shown on the "Tree Diagram" included in this report. Specific methods were tabulated for each secondary function on the Functional Analysis Worksheet.

Many methods were abandoned as being unfeasible or impractical for this building and its program as presently developed and some were impossible to evaluate due to lack of information on the building construction and exact location.

Those systems that were left as feasible were factored and graded as to their predictable success in meeting the "Fire Safety Criteria."

FIGURE A-3
Value-engineering proposal summary for fire safety systems.

CONCLUSIONS

Specifications of Function

    A. Protect Life: No loss of life

    B. Protect Property: 5000 SF max area of loss

(A) Protect Life		(B) Protect Property	
System	Cost	System	Cost
Stairs	452,000	Own fire dept.	3,500,000
Elevators	534,000	Complete removal	
Own fire dept.	3,500,000	of fuel	8,500,000
Sprinklers	325,000	Sprinklers	325,000
Halon	3,500,000	Halon	3,500,000

RECOMMENDATIONS

Provide sprinklers throughout the office spaces in the building. Further study other areas (file, computer, cafeteria, auditorium, and mechanical).

FIGURE A-3 (Continued)

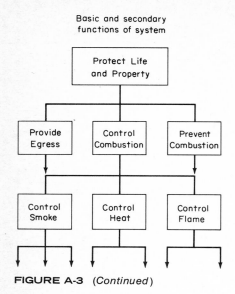

Basic and secondary
functions of system

FIGURE A-3 *(Continued)*

In Table A-1, the creative session brainstorming list is shown. These creative ideas are listed without evaluation, just as they were noted by the meeting scribe.

Figure A-3 presents the summary value-engineering report.

3
THE JOB PLAN

In the words of L. E. Miles: "Value analysis is a system, a complete set of techniques, properly arranged, for the sole purpose of efficiently identifying unnecessary costs before, during or after the fact. Some of the techniques are familiar, some modified, some new. . . . It is a disciplined action system. . . ." (Ref. 1).

The job plan is the disciplined system which combines the special technology of value analysis with other procedures and techniques to result in the complete analysis.

The six basic phases in the job plan as identified by the Department of Defense are:

1. Orientation

2. Information

3. Speculation

4. Analysis

5. Development

6. Presentation and follow-up

Figure 3-1 shows these stages with subheadings posing questions addressed to the item under study. Table 3-1 lists the job-plan descriptions as described by four DOD and GSA handbooks and three experts in the field.

Table 3-1
Value-Engineering Job-Plan Categories

L. D. Miles* (3) 1961	DOD HANDBOOK 5010.8-4 (1963)	DOD (USA META) 1968	E. D. Heller† (1) 1971	A. E. Mudge‡ (2) 1971	GSA-PBS P 8000.1 1972	L. D. Miles§ (4) 1972	PBS VM workbook 1974
Orientation		Orientation		Project selection	Orientation	Information	Information
Information	Information	Information	Information	Information	Information	Analysis	Function
Speculation	Speculation	Speculation	Creative	Function	Speculation	Creation	Creative
Analysis	Analysis	Analysis	Evaluation	Creation	Analysis	Judgment	Judicial
Program planning	Development	Development	Investigation	Evaluation	Development	Development	Development
Program execution	Presentation	Presentation and follow-up	Reporting	Investigation	Presentation		Presentation
Summary and conclusion			Implementation	Recommendation	Implementation		Implementation
					Follow-up		Follow-up

* L. D. Miles, *Techniques of Value Analysis and Engineering*, 1st ed., McGraw-Hill, New York, 1961.
† E. D. Heller, *Value Management: Value Engineering and Cost Reduction:* Addison-Wesley, Reading, Mass., 1971.
‡ Arthur E. Mudge, *Value Engineering*, McGraw-Hill, New York, 1971.
§ L. D. Miles, *Techniques of Value Analysis and Engineering;* 2d ed., McGraw-Hill, New York, 1972.

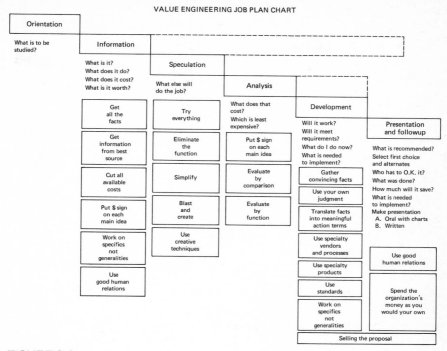

FIGURE 3-1
Value-engineering job-plan chart. (U.S. Department of Defense Handbook. 5010.8-4, *Value Engineering*.)

ORIENTATION

Orientation actually precedes the development of the job plan, and so it has been omitted as the first phase of job planning in many considerations. To orient is to "place or arrange relative to an external reference frame. . . ." In a properly established value-analysis situation, this would include the selection of appropriate areas to be studied and the appropriate team to accomplish the study. In an organization which has been conducting value analysis, policies will have been established which assist in these determinations.

As illustrated schematically in Fig. 3-2, the greatest potential for net cost savings occurs in the earlier stages of a project so that, given a choice, the value-analysis team would select study areas in projects in the earlier stages of the project cycle. Naturally, this is controlled by the nature and number of projects which can be considered for study. If the organization has under its control only projects which are in construction, these represent the only area of opportunity. Conversely,

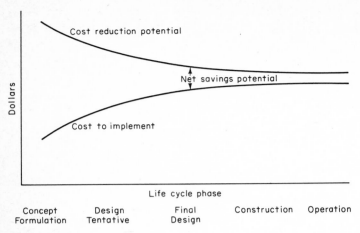

FIGURE 3-2
Potential for net cost savings. (Public Building Service/GSA, Manual P 8000.1, *Value Engineering.*)

the question may not even be one of selecting projects. If, in fact, the organization has one major project, the value-analysis team has the initial assignment of orienting its studies to the area of greatest potential return.

INFORMATION PHASE

In the words of TV's Sergeant Joe Friday: "Just give us the facts. . . ." The information phase is a fact-finding phase. The purpose is to accumulate all the factual information available in regard to the proposed area of study. If the information phase is the initial one, this may require collection of information on the entire project so that the selection of the most opportune area for value analysis can be made. If this area has already been selected, the factual collection can focus upon that one phase of the project for this sector of the value-analysis study.

In a construction project, the facts available will depend to a great degree upon the phase of design or construction. Collection of the facts should include assembly of the best available program parameters, design specifications, and flow diagrams and/or drawings available at the time.

Background information should be assembled and subdivided into facts and assumptions.

The information phase is not one of evaluating or judging. The maximum amount of pertinent information should be gathered in the minimum amount of time.

In addition to facts and assumptions, any information required for further analysis but not available should be delineated. The team should sort out extraneous matter in preparation for the analytical phase.

FUNCTIONAL ANALYSIS

This step is included as part of the information phase in many job plans (see Table 3-1). However, it is the very heart of the value-analysis effort, and both Miles and Mudge, two of the leading experts, single out the functional analysis as a separate phase. When this is not done, the functional analysis would necessarily be the latter portion of the information phase.

The key value questions are addressed to the area under study at this point:

What is it?

What does it do?

What does it cost?

What is it worth?

The function of the key area under study is described in a two-word phrase, one noun and one verb. Any number of secondary functions may also be described.

CREATIVE-SPECULATIVE PHASE

The purpose of this phase is the generation of ideas for alternative solutions to the function being analyzed. The emphasis is upon *imagination*, and free-form think-tank approaches or brainstorming are very successful. Miles suggests a group of 3 to 10 people as the optimum.

The purpose is to generate ideas—not to judge them. Nothing should be done to stifle a creative process, for the thought-association process may generate feasible ideas in response to abstract solutions which may seem ridiculous when listed.

JUDICIAL ANALYSIS PHASE

The ideas developed in the creative phase are screened to eliminate those which do not meet the functional requirements. The viable alternatives are identified and ranked according to feasibility and cost. Various techniques can be used, most based on a matrix comparison, to develop and select the best solution.

DEVELOPMENT PHASE

The alternatives selected for potential implementation are now reviewed in depth. The ideas are checked out with purchasing departments, vendors, and other specialists in the field. Cost factors are verified. Ideas are reviewed for flaws or problem areas, and a development plan is established. The plan is based upon the alternative or alternatives which are economically viable.

SUMMARY

Figure 3-3 is a summary of the major phases of the value-engineering job plan as described by PBS/GSA Handbook P8000.1. It summarizes the thinking processes involved in the value analysis of a project or components of a project. It should also serve to illustrate the difference between the typical analysis of systems during design and a complete value analysis according to a value job plan.

Following the job plan approach organizes the value analysis into distinctive stages. This permits the individuals to assume different viewpoints, separating, in particular, imaginative speculation and creative thinking from evaluation and judgment.

In a note of counterpoint, a reviewer in commenting on the thoroughness of treatment of the value-engineering job plan by Mudge pointed out that the rigorous job plan should not preclude or ignore "quantum jumps," which on occasion occur. In fact, many situations pointed out as value judgment occur without the benefit of a rigorous job plan—probably in spite of the lack of a plan rather than because of it. One such quantum jump would have to be the New York Port Authority decision to collect one toll for all vehicles entering New York, virtually cutting in half the cost of toll collection, and yet this policy change came more than 30 years after the first tunnel was put into operation. In Ref. 2, Miles describes one such quantum jump in value analysis. In the process of defining the function for a concrete wall, a construction engineer came up with a rapid value analysis and much improved solution.

The situation occurred in the construction of a new laboratory which would contain powerful x-ray equipment used to inspect large

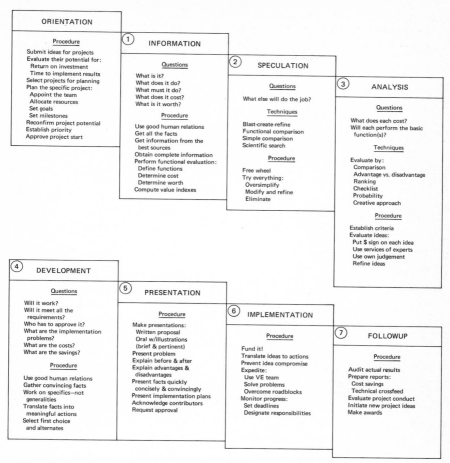

FIGURE 3-3

Major phases of the job plan. (Public Building Service/GSA, Manual P 8000.1, *Value Engineering.*)

forgings. For radiation protection, the drawings included a concrete wall 7 ft thick by 14 ft high in the shape of a horseshoe. The bid cost for this wall was $50,000, and work would have proceeded if the plant manager had not placed a condition upon the laboratory manager that the removal of the wall in the case of relocation of the laboratory would be the laboratory manager's responsibility.

At this point, the laboratory manager reconsidered the wall and asked the advice of the construction engineer. This individual had value-analysis function-based experience and recognized that the basic function of the wall was "stop radiation." Investigation indicated that dirt would also provide the same basic function, and that a mound 14 ft high by 14 ft thick would perform the same basic function and could be installed for $5,000. The alternative was immediately approved.

BLAST-CREATE-REFINE

In Ref. 3, a special technique is described which can be used to reach value objectives in specific problems. The technique is carried out in the following stages:

1. **Blast** As described in the DOD manual, blast means to get off the beaten path. Basic functions are identified, but alternatives which are suggested may cover only part of the function. The approach may be traumatic to the originator of a design or plan, but it tends to break away from the status quo or inertia

2. **Create** Creativity as described by the DOD is reaching for an unusual idea or a totally new approach. As described by Miles, it is the use of intense creativity to generate alternative means by which concepts revealed by the blasting can be modified to accomplish a large part of the function.

3. **Refine** In the DOD parlance, refining involves strengthening and expanding ideas which suggest a different way to perform the function. Miles suggests that the necessary created alternatives be added to the functions which would be accomplished by the blasted product and the total sifted and refined until the refined product fully accomplishes the total function.

The blast-create-refine technique incorporates all the methods and procedures of value analysis but focuses them more specifically on a single area and uses a team approach to generate results perhaps more rapidly. The blast-create-refine approach should be applied with judgment. The very term "blast" suggests a brute-force breakthrough of roadblocks and inertia. Necessarily, negative feelings will be set up either with part of the organizational structure or within the team itself. Accordingly, the organizational groundwork has to be properly laid when this high-powered focused type of value analysis is applied. Further, the "blast" approach applied without an overview consideration of the power factors within an organization can result in the winning of a battle and the loss of a war. Where the value-analysis purpose can be achieved without the blast stage, the normal routine of value analysis should be used. Using an elephant gun to kill an ant is an overinvestment—and poor value.

REFERENCES

1. Miles, L. D., *Techniques of Value Analysis and Engineering,* 2d. ed., McGraw-Hill, New York, 1972, p. xvi.

2. Ibid. p. 26.

3. Ibid, p. 102.

CASE STUDY B
THE JOB PLAN

One of the teams conducting value analysis in the predesign stage of the SSA payment centers evaluated the various possible locations for a planned cafeteria.

ORIENTATION

Orientation and selection of the parameters of the analysis were done prior to the value-analysis session. The value-analysis team was given the following orientation instruction regarding design criteria.

"The facilities should include a cafeteria adequate to accommodate SSA employees considering staggered lunch periods. The space needs in net square feet are 15,000 to 16,500 square feet."

"Cafeterias shall include a food preparation area, offices and restrooms for cafeteria employees, serving lines, and an eating area in accordance with the number of persons employed. The layout in decor shall be subject to approval of SSA. The cafeteria shall be equipped in accordance with PPBS P5800, 18A, Chapter 17, Concessions. The size of the cafeteria is detailed in Tables 1 through 4."

Design History and Background The criteria stated above were taken directly from SSA criteria. The following alternatives were to be considered along with any others that the team might generate.

1. Located on the top floor
2. Located on the ground floor
3. Located on the middle floor
4. Located in a separate building
5. Located in a separate building combined with the auditorium
6. Actual need for a cafeteria

INFORMATION PHASE

The team undertook a functional analysis of the cafeteria based on the criteria. They viewed the cafeteria as having two distinctly separate functions, one as an assembly area and the other as a food-handling area. Figure B-1 shows the results of the functional analysis.

In the speculation phase, the team used the brainstorming method to prepare a list of approaches Table B-1.

PROJECT _____ NAME _____

ITEM *Cafeteria* _____ TEL. NO. _____

BASIC FUNCTION _____ DATE _____

QUAN TITY	UNIT	ELEMENT DESCRIPTION	FUNCTION VERB	NOUN	KIND	EXPLANATION	WORTH	COST
		Food handling area	Facilitate	Food handling	B			
			Receive	Materials	S	All supplies		
			Store	Materials	S	"		
			Process	Food	S			
			Dispence	Food	S			
			Dispose	Waste	S			
			Provide	Cleanup	S			
			Provide	Office space	S			
			Provide	Toilet space	S	Employees		
			Provide	Locker space		"		

(a)

PROJECT _____ NAME _____

ITEM *Cafeteria* _____ TEL. NO. _____

BASIC FUNCTION _____ DATE _____

QUAN TITY	UNIT	ELEMENT DESCRIPTION	FUNCTION VERB	NOUN	KIND	EXPLANATION	WORTH	COST
		Assembly area	Assemble	People	B	For eating		
			Provide	Social contact	S			
			Provide	Environment change	S			
			Controls	Movement	S			
			Provide	Circulation	S	Space rqmt.		
			Provide	Eating facilities	S	Furniture		
			Assemble	People	S	For meeting		

(b)

FIGURE B-1 (a) and (b)
Results of analysis of cafeteria functions.

Table B-1
Results of Brainstorming Session to Determine Cafeteria Location

Food-handling facilities

1. Automat
2. Preorder
3. Prepare on site (in-house staff)
4. Cater—prepare on site
5. Cater—prepare off site
6. TV dinner style
7. Hostess serving
8. Self-serving
9. Hostess clean-up/busing
10. Soft clean-up/busing
11. Portion control
12. Piped-in food to various floors from food-dispensing area
13. Dehydrated foods/condensed foods
14. Outside garden—pick your own
15. Self-service soup/sandwich
16. Pay with tickets
17. Pay with Social Security card—deduct from paycheck
18. Prepay meals with standard deduction
19. Intravenous feeding
20. MacDonald-type system
21. Disposal units
22. Garbage cans
23. Compactor with/without pulper
24. Mobile kitchen/complete mobile facility

Assemble people

1. Provide lounging-eating chairs
2. Have TV/audiovisual/educational
3. Overlook gardens/views
4. Design intimate spaces
5. Include related art/handicraft display spaces
6. Consider distributing areas over building and use for fire, share use

| CONSTRUCTION COST ESTIMATE | | DATE PREPARED 15 February 1972 | | SHEET | OF |

PROJECT
GSA/PBS - SSAPC

LOCATION
Add. Cost to Locate Cafe. @Top Flr-10th of

ARCHITECT ENGINEER —
10 Stories

BASIS FOR ESTIMATE
☐ CODE A *(No design completed)*
☐ CODE B *(Preliminary design)*
☐ CODE C *(Final design)*
☐ OTHER *(Specify)* _____

DRAWING NO.   **ESTIMATOR**   **CHECKED BY**

SUMMARY	QUANTITY		LABOR		MATERIAL		TOTAL COST
	NO. UNITS	UNIT MEAS.	PER UNIT	TOTAL	PER UNIT	TOTAL	
Excavation &Foundation							-----
Structural-14' Flr to							
Roof							
Stairwell-Incl Wall,							
Stairs,Finishes 5/Fl	50	EA			8400		420,000
Roofing							-----
Exterior Wall							-----
Interior Finishes							
1.50 + .30							
Demising Walls	5500	SF			180		10,000
HVAC							-----
Plumbing & Sprinkler							-----
Electrical							-----
Elevators							-----
General Conditions	8.7%						37,400
							467,400
						Use	468,000

NOTE: Possibly additional costs in piping to 10th floor for food handling equipment. Additional use put on elevators to accommodate food handling supplies.

ENG FORM 150
1 AUG 59   PREVIOUS EDITION MAY BE USED   * U.S. GOVERNMENT PRINTING OFFICE : 1959 O—516148

FIGURE B-2
Detailed estimate of incremental cost of one of five alternative approaches.

VALUE ENGINEERING PROPOSAL SUMMARY		DATE
		18 February 1972

TO:	FROM:	PHONE NO
GSA/PBS	Turner Construction Company Value Engineering Team No. 3	

ITEM NAME		
Cafeteria		

COMPONENT OF	QUANTITY	STUDY SPAN:
SSA Payment Center	1 per bldg.	START 2/14/72
		COMPLETE 2/18/72

FUNCTION OF ITEM:		ESTIMATED SAVINGS:
VERB	NOUN	
Facilitate	Food Handling	

DESCRIPTION OF PRESENT DESIGN

Area Requirements Projected to 1975:

```
1)  Cafeteria, kitchen concession office    12,000
    Refreshment rms @ 1500 sq ft each        3,000
    Vending rms. @ 250 sq ft each            1,250
                                            16,250 sq ft
```

The kitchen facility is to be capable of complete food preparation, cooking, and serving for breakfast and lunch. The total food service and eating facility will be capable of feeding 2,077 personnel, with each turnover capable of being fed within the 1/2 hr time allowed. The Food service/eating facility shall be capable of future expansion of popualtion of approx. 1,100 people (in 200,000 net usable space). The facility is intended to provide a variety of menu selection and quality food attractive to SSA personnel at an affordable price. The eating space(s) will provide all amenities possible within the budget to improve employee's well-being. Supplementary refreshment and vending areas provided will be considered in their relationship to the cafeteria facility.

CONCLUSIONS

Provide a food management facility which can be operated by a concessionaire and of a type which encompasses partial offsite and partial onsite food prepatation.
The consideration of combining the eating area with the meeting area of the 'auditorium' function should be a topic for further study.

RECOMMENDATIONS

1. Locate facility at ground floor level.
2. Consider combined use of space with auditorium.
3. Use catering service or concessionaire for food management.
4. Study kitchen equipment requirements and storage facilities.
5. Avoid serving-line layout.
6. Provide an attractive seating area with a change in environment.

FIGURE B-3

Value-engineering proposal summary for cafeteria development plan.

ANALYSIS

The key to the analysis of the various approaches was cost. Figure B-2 shows the detailed estimate for the incremental cost for one of the five alternative approaches. The costs are those above the cost of the lowest alternative, which would be the cost of the cafeteria located on the first floor. The comparative costs were as follows:

1. Cafeteria in building on first floor   base cost
2. In building, tenth floor—base cost   +$230,000
3. In building, top floor—base cost   +$470,000
4. Penthouse—base cost   +$770,000
5. Separate building—base cost   +$225,000
    +land cost

In addition to the cost of the facility, three approaches to furnishing food were evaluated. These were:

1. Conventional—in-house operated: The advantage of this approach is direct quality and quantity control. The disadvantage is the requirement that food-processing personnel must be hired and that facility space requirements are substantially greater.

2. Complete preparation off-site—catered: The advantage of this approach is minimization of kitchen storage space and preparation space. Also, the catering service can be replaced if satisfactory quality or prices are not achieved. The disadvantage of this approach is poor quality of food flavor and little individual control of food-portion size.

3. Combination approach: In this approach, there is partial preparation on site to improve quality but continuation of the catered approach for most of the food. This continues the advantage of minimizing kitchen and storage space and also the ability to replace the caterer with a minimum of disruption. A principal disadvantage is poor control of food-portion size.

DEVELOPMENT

The team combined the various prior phases into an evaluation and development plan as described in Fig. B-3, which is a summary of the team's conclusions.

PRESENTATION AND FOLLOW-UP

The team prepared a verbal presentation which included flip charts to describe the results of their study (Fig. B-4).

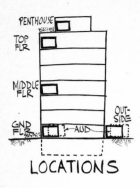

LOCATIONS

PENTHOUSE
TOP FLR
MIDDLE FLR
OUTSIDE
GND FLR — AUD

Creative Session

IDEAS

FOOD HANDLING

· INTRAVENOUS
· DEHYDRATED
· IN FLIGHT HOSTESS
· PIPED IN
· PIPED IN BY CATERERS' VANS

Items for
· FUNCTIONAL CONSIDERATION

FOOD HANDLING	SEATING AREA
·RECEIVING	· EATING
·STORAGE	· MEETING
· PRE-PREP	
·COOKING	
·SERVING →	
· CLEAN-UP	
· REFUSE DISPOSAL	

CAFE LOCATION
INITIAL COSTS

- GROUND FL.   X$ + $ 0.
- MIDDLE FL.   X$ + $230M
- TOP FLOOR   X$ + $ 470M
- PENTHOUSE   X$ + $770M
- OUTSIDE BUILDING   X$ + $225M + LAND COST

KEY IDEAS

A- Conventional
IN-HOUSE OPERATED
DIRECT CONTROL
+ · QUALITY
+ · QUANTITY
− MAXIMUM KITCHEN STORAGE
− EMPLOYMENT OF FOOD PROCESSING PERSONNEL

B- Complete Preparation
OFF-SITE CATERING.
+ MINIMIZE OR ELIMINATE KITCHEN & STORAGE
+ CATERING EMPLOYMENT REPLACEMENT
− FOOD FLAVOR PORTION QUANTITIES

C- PARTIAL PREP. OFF/ON SITE CATERED
+ MINIMIZE KITCHEN & STORAGE CATERING SERVICE REPLACEMENT
− PORTION SIZE

Recommendations

● LOCATION — GROUND FLOOR

● ATTRACTIVE SEATING AREA

● FOOD HANDLING —
A- CATERED SERVICE
B- STUDY KITCHEN EQUIPMENT & STORAGE FACILITIES
C- ELIMINATE SERVING-LINE APPROACH

FIGURE B-4
Flip charts included in value-analysis team's verbal presentation.

4
TECHNOLOGY

The construction industry often appraises itself as old-fashioned and slow to evolve technologically. Nevertheless, evolution has been taking place, and better products and procedures are available, some on a grand scale.

The creative phase of value analysis should not be limited by the technology available, but most certainly the value-engineering team should be knowledgeable in terms of the materials, techniques, and new ideas which have been tested. This permits the creativity to start from a much higher datum point or level.

Conversely, the value analyst should avoid acceptance of technology for technology's sake. For instance, in Operation Breakthrough, sponsored by HUD, an attempt was made to develop building systems which would be economical. Many facets of the overall program were successful, but in almost every case standard stick housing could be brought in at equal value for a lower cost than the building systems. The lesson is clear—the value analyst cannot assume that technology is value effective.

In aerospace at Cape Canaveral, a common jest was the statement that an elephant is really a mouse built by NASA. The NASA mouse did the job, however; it got to the moon. The man-on-the-moon program did stimulate many industries, including the construction industry. In many cases, existing technology was utilized to the ultimate to deliver massive facilities on a tight timetable. In other cases, new materials and concepts were developed and implemented. These advances have been integrated into the construction industry and form part of the background of construction technology. It is part of the role of the value analyst to assist in the selection of which technological improvement should be used and at what point or specific situation within a project.

Technology is oriented toward the improvement of construction materials, construction methods, or construction equipment. The value impact has to be considered in terms of a total equation. For instance, a more expensive material may be less expensive in terms of field personnel, with the total effect of saving installed costs. Some materials may have a longer life at lower maintenance cost, resulting in a long-term life-cycle saving. Plastic pipe properly utilized is a technological improvement which saves first cost and life-cycle cost, resulting in a definite value improvement. Unfortunately, many building codes continue to block its utilization. The value analyst has to be aware of those technological improvements which can reasonably be utilized.

MATERIALS

In Ref. 1, R. G. Zilly questions "whether new design concepts create new materials or whether new materials create new design concepts. Only recently, however, has construction technology occasionally taken the dominant role and preceded introduction of new materials and new design concepts."

Technical and engineering periodicals are constantly reviewing new materials and projecting their potential impact. This coverage includes not only professional reports but also proprietary advertising by manufacturers.

Concrete

Cement-based concrete is one of the oldest building materials. One of the greatest breakthroughs in the evolution of the construction industry was the development of reinforced concrete, with its ability to carry tensile loads. Improved batching and mixing equipment makes higher-strength concrete more feasible and economical. Whereas 3,000 lb/sq in. concrete was once considered the top quality range, today's buildings are often using 5,000 lb/sq in. and even higher strengths. The result is smaller sections which require less time for handling and provide more net usable cube space.

Continued research on methods of handling concrete, use of special air-entrainment mixes, better control of aggregates, and skillful use of admixtures all contribute to the continuing improvement of concrete. Experiments in more durable types of reinforcement, such as glass fibers, offer a potential for even more improvement in the basic material.

The equipment for handling concrete has also improved. The Georgia buggy and crane bucket are being replaced by the concrete pump and conveyor. The development of many new forming systems and the bond breakers which help make them more practical add to the capabilities of concrete design.

Slip forming, once limited to core structures and stacks, is now being used for shear-resistant portions of the structure and special adaptations. Properly utilized, slip forming provides high value at relatively low unit costs.

Precast concrete is being widely utilized for exterior panels and structural members. Prestressed and posttensioned components are being used for shear-resistant portions of the structure and special prestressed-concrete members and panels can provide high-quality results at low unit cost. Conversely, each application must be evaluated, and in many situations, cast-in-place concrete is more competitive in terms of value in place.

Wood

Another of the oldest construction materials continues to evolve in its ability to provide better value performance in construction. Laminated wood members are the result of the combination of equipment and technology involving synthetic resins. Plywood has been an important factor in concrete forming and continues to show its versatility in textured finishes as well as special finishes and paneling for interior work.

Heavy timber construction has been shown to be fireproof if properly designed and shielded, and wood continues to show its versatility in both standard and prefabricated housing.

Wood can be flame-treated to make it more fire-resistant, and sheathed in fireproof gypsum board, wood stud partitions can meet code fire-rating conditions. Continuing advantages of wood are its easy workability and ready supply in standard stress-graded sizes.

Metals

Structural steel has evolved slowly but firmly over the past few decades. A large portion of this development has been in the improvement in connections, with high-strength bolts virtually replacing riveting. This has improved productivity and produced a better product. The steel itself is available in higher strengths, permitting smaller cross sections to perform the same load requirements. This provides better net usable cubage in a building. New shapes and cross sections are being used more frequently. "Corten" weathering steel properly utilized can substantially reduce life-cycle costs for maintenance.

Structural metals are not the only ones to improve. Lightweight joists are available for purlins and interior framing members, and even in some cases, as main framing members on single-story buildings. Metal roof decking and floor decking can be placed rapidly.

Steel and aluminum panels have become highly developed recently, particularly for cladding the exterior of high-rise buildings. Metal and hollow metal frames and bucks are compatible with high productivity and low life-cycle cost. The availability of exotic metals developed in the aerospace field is becoming increasingly important in the design of mechanical and electrical equipment. Welding techniques have been improved, thus broadening the use of aluminum and stainless steel.

Plastics

The postwar era has been one of rapid development in plastics, and many have taken an important place in the construction industry. Foamed and preformed plastics are important insulation media. Acrylics provide clear plate plastic for glazing and are the vehicle for a large portion of the paint used today.

Plastics and derivatives from plastic technology have provided a wide range of new caulking materials and sealants.

The adaptability of plastics to be injection-molded has provided a relatively low-cost availability of high-quality shapes and forms of wide utility.

Plastic laminates provide kitchen and bathroom cabinetry with highly durable surfaces at relatively low cost.

Other materials such as glass, masonry, aluminum, bituminous products, rubber, fibers, insulation, and innumerable others continue to evolve through applied research conducted by industry groups, user's organizations, and manufacturers themselves. The key to the ultimate value of the material lies in the careful selection of its application.

Space age adhesives, coatings, and fasteners, in particular epoxies, have great impact upon the installation of ductwork, piping, ceilings, tile, and many other interior items. While most do not change the basic sequence of installation, they are fast and convenient (when properly selected and utilized).

EQUIPMENT

There has been a continual improvement in construction equipment. The WABCO Construction & Mining Equipment group has published a series of advertisements documenting this evolution. Figure 4-1 shows the evolution of the grader from 1885, when the "Little Wonder" J. D. Adams grader evolved from the plow into the first leaning-wheel grader. Also shown is a vintage 1900 improvement and the first all-welded grader, which dates from 1930.

Figure 4-2 shows the evolution from horse-drawn truck to the first scraper, which was built by equipment pioneer R. G. LeTourneau. Also shown is LeTourneau's first scraper with electric controls and, developed in 1938, the first Tournapull rubber-tire grader.

Highway paving equipment has progressed from fixed forms and a series of individual screeding and finishing machines to an integrated equipment train which batches, mixes, places, and finishes paving in a continuous slip form technique. Similar but special continuous-placing equipment has been developed for canals and waterways.

Tunnels are drilled with specially designed moles which can provide a finished surface faster, more safely, and at a lower cost than many of the old placement techniques. Value analysis in the highway, bridge, and tunnel industries in particular led to the development of special equipment for special projects. Equipment is often amortized on a single project. While in most cases value analysis leads to the development of effective special equipment which makes a particular design or contractor competitive, in some situations the reverse is true. In *Civil Engineering* (Ref. 2), there is a description of a highway contractor who, when his volume of highway contracts was reduced, developed methods of utilizing his highway pavers to place slab-on-grade for industrial and warehouse buildings. He next placed lift slabs on top of the slab-on-grade, using that for a form. Doorways and windows were cut with masonry saws, and the specially reinforced lift slabs were tilted into place at a very low cost.

Building projects tend to make less intensive use of heavy and special equipment, although cranes and hoists are an indispensable part of multistory construction—particularly high rise. A variety of tools and special equipment have been developed to improve the productivity of the individual worker. These include hand tools and power-actuated tools as well as special hoists and lifts for high bay work in a single-story building.

In *Contractor's Management Handbook* (Ref. 1), Zilly suggests that many of the operations in a construction project are actually materials-handling problems. On that basis, he recommends the use

(a)

(b)

(c)

FIGURE 4-1
Evolution of the grader from 1885: (a) "Little Wonder,"
first leaning-wheel grader; (b) improved grader; (c) first
all-welded grader. (Courtesy of WABCO Construction &
Mining Equipment.)

FIGURE 4-2
(a) through (f) Evolution from the horse-drawn truck to the first scraper. (Courtesy of WABCO Construction & Mining Equipment.)

of the construction equipment in the following cases:

1. Loads which must be lifted from floor level to overhead level frequently

2. Loads which require more than one man to lift them

3. Loads which are 75 lb or more and must be lifted above a man's knees

4. Loads which must be handled continuously for over 30 minutes when a man is normally assigned to tasks other than lifting

5. Loads which must be lifted for long periods of time

6. Loads which must be moved by one or more men for horizontal distances of over 5 ft

Cranes and other lifting equipment have increased in size, capacity, and versatility so that even with increase in costs there has been a substantial increase in units of production per dollar of capital invested. High-capacity cranes have been made more mobile by the introduction of 360-degree handling capability—as in the Manitowoc Ringer. Previously, only lightweight cranes had rubber-tire mobility, but now the medium range provides this for better mobility on the job site with low earth-bearing pressures. Special climbing cranes such as the tower cranes provide versatile lifting from a fixed point, such as elevator shafts or chimneys, appearing to literally lift themselves by their own bootstraps.

DESIGN CONCEPTS

New concepts in design may result from technological advances in materials or equipment or may be a breakthrough in the application of existing materials and equipment.

Good design is the careful blending of new and old technology to produce the best value for the design problem to be solved. Unfortunately, new technology is sometimes applied simply because it is new—not necessarily because it provides a better value. Value analysis in technological development is only a sound companion set if the ultimate result produces a better value for the owner.

Foundations

The design of foundations is largely an empirical approach. The key to better value in foundation design lies in better subsurface investigation before design and the use of computers to provide more definitive correlation of field testing with actual design. The field testing is not limited to determination of prior conditions but also should include comprehensive actual field testing.

In the New York City area, the Port Authority has done much in surcharge as a design approach to preload soils to make them suitable for foundations. While generally utilized for large areas such as runways, the Port Authority has also used it in preparing warehousing areas and other structures to accept design loading.

There has been a definite trend for designers to look at a total site when designing foundations. In one situation, the building involved a substantial excavation. Before the excavation was made, footings were drilled down through the overburden and an elevated floor slab was poured with the existing grade used as a support.

Water-table uplift can be a problem, and innovative approaches can be used to counter it. The old answer was deadweight, and perhaps uplift piling. In other sites, it has been possible to literally float a foundation on unstable soil by removing a volume equal to the weight of the contemplated structure so that a balance is achieved. Thoughtful design concepts can provide better protection against seismic action at lowest cost.

Structures

Design concepts in structural design have evolved as a result of the availability of better materials and better connection techniques. Again, the total viewpoint has resulted in the concept of a building as a tube rather than as a combination of beams and columns. The impact has been to take full credit for many of the secondary advantages offered by certain design approaches.

Many of the breakthrough structures, such as the World Trade structures in New York City, demand state-of-the-art design concepts. These, in turn, can be used in basic structures.

HVAC

Heating, ventilating, and air conditioning design concepts have evolved very rapidly under the motivation of the energy shortage. Systems with a higher initial cost can now be considered, since there is wider recognition of life-cycle costing.

New concepts include the use of energy-recovery wheels to conserve energy which had previously been wasted in air-discharge systems. The use of pumps to recover and save energy has been implemented in many recent projects. Under this system, the water loop interconnecting the HVAC system is used to capture heat, circulate it, store it, or reject it—shifting the energy on a multizone basis to points of different needs. Under this concept, rather than the heating and cooling systems running in opposition to each other, heat removed by the air conditioning system on the sunny side of the building would be used to warm chilled areas on the shadow side. Conversely, energy from lighting and people load would be used during winter weather to provide heat. The system is operated by reverse-cycle heat pumps and applies to a variety of HVAC basic-equipment configurations. Increased use of the concept has resulted in the design availability of double-bundle condensers, although the concept does not demand them.

The air-distribution systems for HVAC have dramatically changed. The standard low-velocity constant-volume duct has given way to dual-duct systems and variable-air-volume (V.A.V.) systems. S. Darzanani (Ref. 3) points out that variable air volume is actually an old concept but little used because of the inability to take advantage of the lower energy required to operate equipment. The author indicates that practical problems associated with implementation of the reduced energy requirement under the variable-air-volume approach have been solved and a number of large-scale applications have been made. (One of these is the new police headquarters in New York City.)

The author claims the advantages of the all-air approach for V.A.V.—principally that air-water units are not required at the perimeter or the individual room. In addition, he indicates that lower initial costs can be achieved because of reduced fan sizes and smaller ductwork.

Constant-air-volume systems achieve no energy savings when working at part load. One of the prime energy- and cost-saving attributes of the V.A.V. is its ability to operate only that portion of the system which is needed. However, in Ref. 4, I. A. Naman, another engineer, points out that in his opinion the V.A.V. promise is largely undelivered. He questions whether there is really a saving in fan horsepower in comparison with constant-volume low velocity—recognizing that the V.A.V. operates at higher velocities and therefore higher energy levels. With lights on and people present, the air conditioning load requires an almost constant-volume operation, particularly in the interior zones. Further, in order to make the V.A.V. system work in the heating phase, it is often necessary to provide a perimeter heating system. Control must be achieved by the use of the V.A.V. cooling load, which amounts to reheating and therefore upsets the saving aspects.

Examination of the new design concepts in HVAC demonstrates that there are many approaches which can work and that value analysis can be particularly useful in the encouragement of a thorough and critical analysis of all factors involved in the total system in order to deliver the best value possible.

One detailed presentation of a closed-loop package heating-cooling system suggests that the system will save 25 percent in energy costs. (As in all such claims, the base question is what is the comparable worth factor.) This approach is a water-loop system connected with each unit in the system to transfer heat energy using reverse-cycle heat pumps. The presentation suggests that the installed cost of this system is less than $3/sq ft with an energy cost of $0.14/sq ft versus fan-coil units at $0.18, central V.A.V. at $0.20, and dual-duct at $0.30.

Electrical Systems

Electrical equipment continues to evolve through a continuing improvement cycle. Electronic control systems have been added to basic electric power controls to provide more reliable and less costly equipment.

Concepts of total energy and selective energy sources can add substantial amounts of value by the reduction of installed cost with no increase in operating cost.

Ross and Kern (Ref. 5) report on a study performed for GSA on requirements for lighting levels. The presentation suggests that the brute-force approach to illumination has gone too far—both in terms of energy requirements and optimal impact upon users. The presentation of results is chiefly concerned with value and suggests that new approaches could potentially save as much as 35 percent of the energy presently used for lighting in what is considered to be optimum methodology. The author suggests a cost ratio of 2.5× present design costs for lighting, which could be recovered at least 35 times during the life cycle of the lighting fixtures. Further, there can be additional potential savings in the initial costs for both electrical and HVAC capacity, which would probably more than pay for the additional design costs.

It is obvious that design technology can be a siren song to the inexperienced value analyst. It is necessary to establish a total cost model for each of the building systems and to evaluate first cost and life-cycle cost for the system, as well as impact upon all other systems, in order to arrive at best value selections.

THE SYSTEMS APPROACH

The value analyst needs a systematic method of introducing new design concepts into the design and construction equation. A low-cost HVAC system may have secondary cost savings in building space and/or HVAC energy equipment, but life-cycle costs may be prohibitive.

A comprehensive cost model can be an effective tool in the reduction of any confusion—and a very compatible adjunct to value analysis.

There have been many applications of systematic approaches to the management of projects, particularly in the aerospace industry. Some of this systemization has been transferred directly to construction, since construction is part of the overall picture in weapon systems and aerospace development.

The senior editor of *Architectural Record* has noted (Ref. 6): "A principal objective of any demonstration program obviously has to be to prove that the project can be accomplished in the first place to establish credibility; further, to insure the costs come in as projected. This can result in sacrifices in quality (particularly appearance) to prove a point . . . Because of the mystique, vagueness and promise, systems projects have been a refuge for clients who cannot or will not analyze their building-related problems and take responsibility for their decisions."

Each and any of the concepts and techniques developed to provide a more systematic approach to the design and construction of facilities can and should be utilized in value analysis. The project manager should select the best applications for the specific project and orchestrate the interrelation between the various concepts and techniques.

Systems Building

Since World War II, the greatest emphasis in technological development in the construction industry has been the effort to develop systems building. Systems building started in the residential field as prefabrication. The basic premise is to utilize industrialized techniques and equipment to reduce building cost. Systems building utilizes building systems in an attempt to produce certain segments or modules of the individual building project in an industrialized environment.

The systems operation may be set up at a major construction site for prefabricating modules on the job site; for the production of housing the concept is that of a production line. This production line utilizes semiskilled workers who utilize heavy specialized equipment such as gang-nailing machines to prefabricate sections of a building.

Basic problems have included the resistance of field craft unions and the size limitations upon modules which can be transported by truck or rail.

Following the gradual demise of most of the major prefabrication efforts in the 1940s, the systems-building approach dwindled to one of almost all talk and no action until the School Construction Systems Development (SCSD) school project in California. The California program established a market of $25 million in school buildings to be bid on the basis of industrialized subsections or subsystems. Manufacturers were invited to develop a product which had to meet certain basic performance specifications. The successful bidder would be guaranteed his share of the initial schools set aside—with the promise of a broad market across the country. The system was perfected, and similar systems were developed in Toronto and Boston. New technology was encouraged, and specifically cost and time control was improved for certain projects. The school-systems approach is being utilized in Broward County in Florida, and it has made a definite contribution. Systems packages have been purchased by the state of New Jersey in the building of several new state campuses. The purchased systems included structural steel, ceilings, partitions, lighting, and HVAC distribution. The major structure and basic HVAC equipment were part of the "out-of-systems" portion.

One of the results of the progress in building systems has been a concerted effort on the part of Building Research Advisory Board (BRAB), supported by many of the federal agencies, to develop a standard set of dimensions for typical building-subsystem modules. The purpose would be to encourage the development of building-systems modules through the guarantee of a broader available market.

One of the problems of building systems is the compatibility between systems. Figure 4-3 shows the basic building-block approach utilized by one of the building systems—the Sectra System. Choice of the structural building system then imposes certain restrictions upon the intersystems as well as HVAC and electrical systems which can be employed. Incompatibility of openings, inserts, and chases can result in an owner's having to pay twice for certain items.

The development of building systems has been along disciplined lines, with the basic systems being principally structural, ceiling, lighting, HVAC distribution, flooring, and partitions. In low-rise buildings, the HVAC package had much more flexibility of development, since it was relatively independent of the structure. In high-rise construction, the preselection of the structural system tends to narrow the flexibility of the development of the other subsystems.

Despite the technological development in building systems, the value analyst must view very carefully the ability of a system to deliver what it has promised. Traditional systems have the advantage of a more proven track record. The editor of *Architectural Record* has spoken with great candor in regard to systems building (Ref. 7):

> There is surely no doubt that the building industry has problems that need solving. While we are charging along at a great rate of production, and producing some very good buildings—residential and non-residential alike—along with some not-so-good buildings, there is no doubt that we need to keep working to cut down the undoubted inefficiency that marks so much building, and get our costs down . . . but I think it's important that we not try to talk ourselves into (or let our clients be talked into) counting on results from "the new systems building" in the absence of any evidence of success.

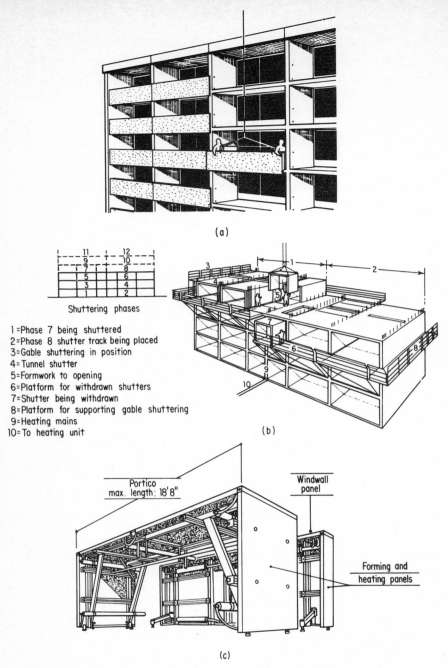

(a)

Shuttering phases

1 = Phase 7 being shuttered
2 = Phase 8 shutter track being placed
3 = Gable shuttering in position
4 = Tunnel shutter
5 = Formwork to opening
6 = Platform for withdrawn shutters
7 = Shutter being withdrawn
8 = Platform for supporting gable shuttering
9 = Heating mains
10 = To heating unit

(b)

Portico
max. length: 18'8"

Windwall
panel

Forming and
heating panels

(c)

FIGURE 4-3
Basic building-block approach. (Sectra System.)

The truth is there has been precious little analysis of the results of any of our major experiments by qualified professionals. There are many efforts underway to import systems techniques that have become popular in Europe, with precious little professional analysis of how effective in cost reduction (or other benefit) they will be in this country.

OPERATION BREAKTHROUGH

One of the best financially supported and well-publicized attacks on technological problems in the housing industry was Operation Breakthrough. In the words of the Housing and Urban Development report on the proposals for Operation Breakthrough (Ref. 8):

> Operation Breakthrough is a broad residential development program designed to resolve a multitude of problems in order to make available quality housing in large quantities. It aims to do this by utilizing modern design and technology, and through contemporary approaches to financing, marketing, land use and management. By helping to create substantial local, regional, and perhaps national markets, Breakthrough is intended to demonstrate the extent of the potential demand—and how it can be aggregated—to those capable of, and interested in, producing both housing and sites for housing. One major objective of the program is to demonstrate that with the kind of large, continuous market enjoyed by the manufacturers of other consumer products, producers of housing in volume can realize economies of scale: they can recover their investments—in research and development, in improvements to their design methods and concepts, and in plants and equipment necessary for volume production.

Breakthrough fell far short of these goals and aspirations. If anything, it proved that for the time being standard, traditional stick housing is more economical and fits the needs and mood of the home-buying market much better than industrialized building-systems approaches. The HUD call for proposals received 244 from organizations willing to design, test, and evaluate complete housing systems and 388 which dealt with concepts. Reference 8 gives in detail 423 proposals released for publication. Over 1,000 firms were involved in Breakthrough, which was initiated on May 8, 1969, but today there are very few that are performing their operations in the manner contemplated at the time of their entry into the Breakthrough concept.

A report by *Engineering News-Record* on the results of Breakthrough (Ref. 9) noted that after 18 months and $30 million in commitments by HUD to builders, not one unit of Operation Breakthrough structure had been started. Operation Breakthrough was credited, however,

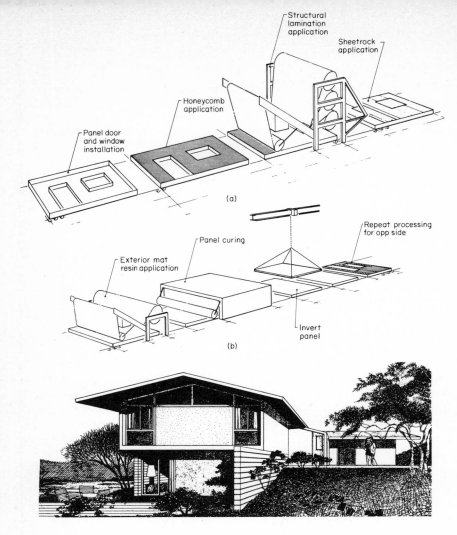

FIGURE 4-4
Aerospace technology in the housing field. (TRW Systems.)

The basic structural sandwich is fabricated from glass fiber, reinforced polyester resin plastic, and the kraftpaper honeycomb, its layers wrapped on large-scale rotary mandrels. The mandrel collapses for stripping the shell. The module's reinforced plastic surface permits mechanical or bonded connections. Cost of the plastic is relatively high, but this is offset by easy fabrication methods. Semiskilled labor is employed in the plant. The kraftpaper is a low-cost item, even after treatment with special flame-retardant chemicals and polymers. In final form the paper is noncombustible as well as moisture-, fungi-, and vermin-resistant. The gypsum board facing provides a fire and heat barrier.

Shells up to 24-ft span have been developed to carry all loads, without partitions, closeout walls, or posts. The basic joining element to tie the modules together is a fiberglass-reinforced plastic U-channel assembly, which provides anchoring for metal fasteners. The shells are crane-lifted onto conventional foundations and attached mechanically or bonded.

58

with focusing attention on the building system. The article quoted an assistant to the president at Lennox, which developed a low-profile rooftop unit in response to the SCSD project: "Many segments of the building industry have lapsed back into their initial, and perhaps more comfortable, method of selling individual products and not entire systems."

In its proposal, TRW Systems, an aerospace firm, combined with Building Systems Development (related to the SCSD program), and some of the largest developers in the country. Their Breakthrough proposal was based upon the manufacture of portable fiber-shell modules wrapped on mandrels. Partitions, panels, and small shapes would be produced in permanent plants and shipped to sites. The system (shown in Fig. 4-4) represents the housing field's utilization of aerospace technology oriented to the development of new materials.

Figure 4-5 illustrates a method which recognizes the problem of coordination and develops a universal building rib. This was proposed by Christian Frey, American Institute of Architects, San Francisco.

Figure 4-6 is the approach recommended by another consortium. It contemplates stacking of modules which can be raised to six stories without extra support. With a separate structural frame support, stacking theoretically could continue to the twenty-fourth story limit.

Figure 4-7 shows two different approaches, each of which is based upon the construction of a central core and then installation of module units from the top down.

Figure 4-8 illustrates that not all the ideas are as outlandish as they might seem. This upside-down building is the 20-story "top-down" building to house 240 families connected with the Soviet Mission to the United Nations. Two core pillars of concrete containing stairs and elevators form the structural frame upon which the balance of the

---

The basic system includes gas-fired forced-air furnaces, adaptable for cooling with the addition of cooling coils to the furnace and condenser units on the roof. Duct runs can penetrate the paper core, terminating at floor-level register-diffusers. Electrical distribution remains outside the structural shell and is surface-applied to minimize problems of design and coordination. Main service to panel boards runs in vertical chases, while wiring for lighting runs at partition-head level in channels that also serve as joint covers and picture-hanger ledges. Several alternatives are offered for the plumbing subsystem, but a standardized layout with vertical lines in continuous chases is proposed for prototype construction.

Innovation land-use concepts in this proposal cover aspects of multiple use, new town development, infill construction and rehabilitation of existing neighborhoods, strategic placement of development, use of floats on water for temporary and permanent (relocation) housing, and preservation of important ecological areas. Open space is a feature of the planning. ("Housing Systems Proposals for Operation Breakthrough," Dec. 1970, U.S. Department of Housing and Urban Development.)

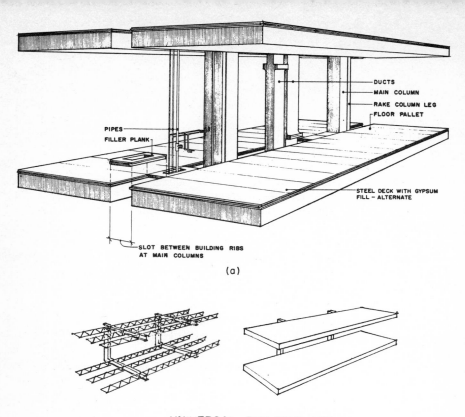

DUCTS
MAIN COLUMN
RAKE COLUMN LEG
FLOOR PALLET

PIPES
FILLER PLANK

STEEL DECK WITH GYPSUM
FILL – ALTERNATE

SLOT BETWEEN BUILDING RIBS
AT MAIN COLUMNS

(a)

UNIVERSAL BUILDING RIB

(b)

FIGURE 4-5

Universal building rib system, Christian Frey, AIA, architect, San Francisco, Calif., proposer.

The basis of the system proposed is a patented, three-dimensional, open-ended, economic, lightweight, load-bearing building component. Designated as a universal building rib, the component is a cantilever, palleted-floor, beam-type unit which utilizes vertical columns, beam arms, and rake-column legs to provide a basic structure. It can be organized vertically as well as horizontally with other like components in chosen mathematical configurations. The system is flexible to the extent that it can provide almost limitless volumetrics, offset levels, and overhangs in a wide variety of architectural configurations. The system is adaptable to and practical for single-family and multifamily low-rise and mid-rise structures. Base materials, beyond the metals used for cantilever beam structure, can be concrete, steel, and aluminum, and gypsum products.

The proposer classifies the total group of primary components in the system in three categories: (1) universal building rib; (2) exterior wall panels (solid, window, and door); and (3) interior wall panels (solid and door). Secondary components include: plumbing, heating and air conditioning, and cabinets and closets. All units lend themselves readily to mass production.

The system is considered to be reduced to practice since similar cantilever structures and secondary applications are almost classic in the construction field, and it is adaptable to central factory or site production. ("Housing Systems Proposals for Operation Breakthrough," HUD.)

60

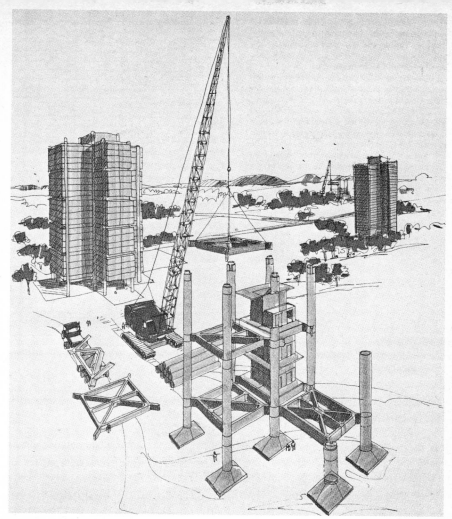

FIGURE 4-6

Stacking modules, National Homes proposer consortium consisting of National Homes Corporation, Housing System Producer, Lafayette, Indiana; Edward Durell Stone & Associates, Architect, New York, New York; Edward D. Stone, Jr. & Associates, Land Planner, Fort Lauderdale, Florida; Semer, White & Jacobsen, Government Relations, Washington, D.C.; Praeger-Kavanagh-Waterbury, Structural Engineers, New York, New York; Cosentini Associates, Mechanical and Electrical Engineers, New York, New York; Computer Applications Inc., Computerized Management System, New York, New York. Factory-produced volumetric modular dwelling units are stacked to six floors without extra support. A separate structural frame support has been developed which permits stacking to a 24-story limit. The proposal also involves extensive use of computer applications to program preparation, site analysis, master plan preparation, design development, and cost-benefit analysis. A principal innovation is use of the vacuum sewage system developed and used in Sweden for the past 10 years, demonstrating a major savings in operating and ancillary construction costs. A volume production of 10,000 units per year is anticipated for the housing system. (Figure 4-6 continued.)

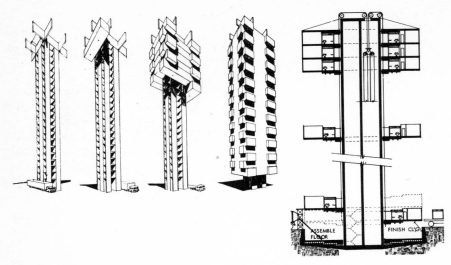

FIGURE 4-7
Two "top-down" concepts. ("Housing Systems Proposals for Operation Breakthrough," HUD.)

building is supported. The living floors are prefabricated at ground level and raised into position.

RESEARCH

The construction industry spends only a limited percentage of its annual business on research. One estimate indicates that the figure is well below 1 percent, which is much less than the industrial average for the United States. Most universities with engineering schools carry on a limited amount of research either as a university function or on

---

Figure 4-6 (*Continued*)

The system embodies four main elements: (1) the three basic planning units—living, utility, and sleeping; (2) the building block assemblies, factory-produced from the basic planning units and delivered to site by truck; (3) the completed dwelling units at the site; and (4) the building itself. All modules are 14 ft wide, affording good room sizes, more house per module, and considerable latitude for architectural arrangement. The building planning system takes full advantage of plant line economies and is not dependent upon specific materials, technologies, or subsystems.

Modular assembly possibilities are extensive. The elements of a dwelling unit are first assembled into the basic modules, then into the modular assemblies (or building blocks), and finally into the dwelling units that compose the finished structure. ("Housing Systems Proposals for Operation Breakthrough," HUD.)

the part of the individual faculty. Prominent in this are Stanford University, Pennsylvania State University, Georgia Institute of Technology, Massachusetts Institute of Technology, and many others. In Ref. 1 the editor notes that the University of New Mexico has begun to compile a directory of information sources for construction technology. The federal government in the course of many of its programs has carried forward construction research, particularly in NASA, the Armed Forces, the Bureau of Reclamation, and the Housing and Urban Development Department. The Corps of Engineers has instituted a number of special research groups. One of these is the Corps' Explosive Excavation Research Laboratory at Livermore, California. Another Corps research center at Vicksburg maintains constant research on the Mississippi River, and at the University of Illinois the Corps has initiated U.S. Army Construction Engineering Research Laboratories, Champaign, Illinois (Dr. L. R. Shaffer, acting director).

FIGURE 4-8
Russian UN mission "top-down" project. (Courtesy of Wide World Photos.)

CONSTRUCTION TECHNIQUES

Technological development is not limited to materials, equipment, and design concepts. Many field-construction techniques either lead to the need for technological development of materials and equipment or optimize the utilization of available materials and equipment.

In *Construction Methods and Equipment* (Ref. 11), articles discuss a list of some 74 construction techniques which are typical of field-construction technological development. Some of these include the following.

Concrete Pumping

Describes the evolution of concrete pumps from specialized equipment for tunnel lining just 10 years ago to highly versatile concrete placers, competitive with pours of all kinds. Pump capacities of up to 125 yd/hr are discussed, and the cost is competitive even at more than $65,000 per unit. Economics depend upon the amount to be pumped, distance of pumping, line size, and line pressure.

Earthmoving Equipment

Close-order scheduling is often the key to the effective utilization of equipment. One contractor on a California interchange utilized smaller equipment (Caterpillar 627 16-yd scrapers) than adjoining contractors who were using a larger Caterpillar 657B, which carried 10 more yards per cycle at a slightly greater speed. The answer to the successful application was the higher maneuverability of the small units and the manning of the machines by operators skilled in double and triple push-pull techniques in which one or more units assist a loaded unit.

A Connecticut contractor working in Florida established a competitive edge which won the project by basing his bid on mucking half a million yards of swampy material rather than depending upon conventional dragline excavation. The production rate was 8 times as fast as the dragline. Since the muckpile ahead of the Allis-Chalmers HD-41 bulldozer was often higher than the dozer, the contractor added a 5-ft extension to the already large 4 × 17 ft blade shields to prevent the radiator from being engulfed by muck.

In Colombia, South America, a contractor, placing an aqueduct which consisted of 79-in. diameter sections weighing 16½ tons and had to traverse a steep hill, built a special rail car and railroad track to handle the aqueduct on the 60-degree inclined portion.

In Texas, a contractor utilizing a Parson's trenching machine to dig through hard soapstone modified the unit by changing the cutting pattern and installing a set of super-hard carbide teeth. The trencher was able to cut 150 ft per day, making cuts 12⅓ ft wide and 27 ft deep, through the flintlike soapstone. Before the modification, progress was only a few feet per day; after the modification the peak production was 240 ft in 10 hours.

Materials

Contractor's value analysis often determines that investment in additional materials, even better than specified, may result in a better product installed at a lower overall cost. One California contractor cut his time and cost of installing 22,000 ft of interceptor sewer line by backfilling the pipe in a soil-cement slurry instead of burying it in compacted sand or gravel. The slurry effectively filled all voids around the pipes, setting up in just ½ hour to firmly lock the buried sections into place. The hardened slurry protected the pipe from any accidental loading such as a cave-in and eliminated the need for shoring of the trench walls.

In the state of Washington, a contractor evaluated the cost of free aggregate available at old aggregate pits and found that purchasing of a higher-grade aggregate would actually save money because of the much lower percentage of waste. The free aggregate was so poor that more than 62 percent of it would have been waste material.

REFERENCES

1. Zilly, R. G., *Contractor's Management Handbook,* McGraw-Hill, New York, 1971, pp. 13–14.

2. *Civil Engineering,* June 1974, p. 58.

3. Darzanani, S., "Design Engineers Guide to Variable Air-Volume Systems," *Actual Specifying Engineer,* July 1974, p. 68.

4. Naman, I. A., "Which Air Conditioning Systems *Really* Save Energy?" *Actual Specifying Engineer,* Aug. 1974, pp. 54–57.

5. Ross, D. K. and G. M. Kern, "New Prescription for Today's Lighting Levels," *Actual Specifying Engineer,* July 1974, p. 79.

6. Fischer, Robert, *Architectural Record,* McGraw-Hill, New York, Oct. 1970.

7. Wagner, Walter F., Jr., *Architectural Record,* McGraw-Hill, New York, Sept. 1973.

8. Department of Housing & Urban Development (HUD), "Housing Systems Proposals for Operation Breakthrough," Government Printing Office, Washington, D.C., Dec. 1970.

9. *Engineering News-Record,* Oct. 22, 1970, p. 31.

10. *Engineering News-Record,* July 4, 1974, p. 12.

11. *Construction Methods and Equipment,* a monthly publication of McGraw-Hill.

In terms of overall time and involvement, the actual construction phase is just the tip of the iceberg of the construction project. There are preconstruction activities which last from 1 to 3 times the actual length of the construction itself. While activities in this preconstruction phase are of substantially lower cost than the actual construction, their impact upon the total cost is substantially greater than the activities of the constructor.

Figure 5-1 illustrates the characteristics of the typical construction project. Some activities in private industry may have more of an overlap, particularly funding. Nevertheless, the overall sequence is generally correct.

The potential for application of value analysis is greatest in the earlier phases, but the information available is necessarily more limited.

The point in the project cycle at which value analysis is applied has a substantial impact upon the range of savings which may be realized. Working with the completed design just prior to the bidding and award process, the value analyzers can, unfortunately, plan to make fewer changes. By this time, the design die is essentially cast. As the project evolves, there is a "funnel effect"; that is, the change level which can be incorporated without substantial time impact continually decreases. The potential value-analysis benefits similarly reduce in range so that from all viewpoints the earlier the value analysis application is made the better.

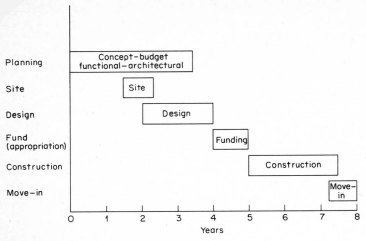

FIGURE 5-1
Sequence of the typical construction project.

PREDESIGN SEQUENCE

Concept Planning

This phase may last from weeks to years. It starts with the definition of the project and its evolution to an authorized project. In government cycles, this can be authorization in the form of an approved budget, while in the private sector it may be the approval of funding sources. Projects may be a continuation of an existing program, as in the case of housing. Other projects evolve almost on a de facto basis, approved finally when the need is so obvious and so strong that no questions can be raised.

In the public sector, the problem of budgeting is one of selecting those projects most needed in an environment of limited resources. In the private sector, the choice is more one of best return on investment.

It is almost axiomatic that management makes conceptual decisions based upon insufficient evidence and information to support a project. Conceptual planning should include a selective process to focus efforts on projects which have a reasonable potential for approval.

Budgeting

Each organization has its own procedures for selecting projects which it determines should be constructed. In the public sector, both government and corporate, there are formally prescribed conditions and procedures to be met in the process. These are usually made functional by an informal set of routine actions which evolve through experience.

Funding

Funding may require separate authorization or may be dependent upon the sale of corporate or government bonds. In the private sector, funding may be concluded by the commitment of a bank line.

Feasibility Studies

As the project takes more definitive form, it may be appropriate to perform a feasibility or economic study of the project and its proposed results. Having passed through the budgeting and planning stages, more specific information will be available in regard to the project. Usually, the more optimistic view which led to the conceptual approval will have been mitigated by actual information, and in the majority of cases the economic viability of the project must be tested. For public projects, economic justification is often difficult because of the intangible socioeconomic values inherent in public works. Nevertheless, many major organizations, including the Corps of Engineers Civil Works Program, must justify the investment by at least a 5 percent annual return.

Of course in the private sector, a much higher return is necessary to have a project go forward. The actual level is usually a function of the business environment, including the returns on investment available in the more conservative bond markets.

The complete feasibility study may well lead into a financial study, including a complete balance sheet of income and outflow of anticipated funding for the project. This plan can be key to the sale of bonds based upon the future revenues of the project. In a recent situation in the Hackensack Meadowlands program, action by New York State in increasing the handle, or portion of the track gate going back to the betting public, mandated a similar consideration in the new Hackensack Meadowlands, shifting its financial study from a positive to marginal situation and therefore inhibiting the sale of the bonds and delaying this major complex.

Programming

Programming is a stage often omitted, and the omission inevitably costs the owner time and money. Programming is usually accomplished in a series starting with the functional program and concluding with the architectural program.

The functional program defines the project in terms of purpose, scope, and functions. Based upon information developed during the planning and budgeting stages, it should set forth clearly the needs to be fulfilled, including the specific services to be provided, and if available, the organizational structure and staffing pattern. The functional program should also set forward proposed operational policies and procedures. Anticipated utilization factors and service loads should be estimated to complete the functional data. The functional program provides the basis for development of the architectural program.

The architectural program is a development from the functional program. Often functional programs are developed by consultants, but the architectural program should be developed either by an architect or a combined team of knowledgeable individuals, including consultants, and architects. The architectural program is a statement of the design problem that the architect is called upon to solve. Unless the functional program is worked out in advance, the architect will be required to make many assumptions or suppositions. The architectural program should state every functional consideration affecting the design. As such, the written architectural program translates needs and objectives into physical requirements. It forms a transition between the statement of broad functional requirements and the details involved in architectural design. Its purpose is to serve as a guide to the architect and is a measure for evaluating his design.

Site Selection

As the program requirements are better defined, the factors required for site selection become more definitive. Site selection will have a strong role in the economic study, and often in the private sector options are placed on proposed sites to preclude escalation of the costs after the location of the project is known.

In site selection, a broad look at environment must be taken, and in many cases an environmental impact report filed with the Federal Environmental Protection Agency as well as, in the case of states such as California, with the state and even local agencies. In any site located in tidewater areas, an environmental impact statement would be filed with the Corps of Engineers.

DESIGN PHASE

The design phase of a project is a complex interplay between the design disciplines, usually under the direction of the architect, particularly in building projects.

Until the industrial revolution, construction was an art performed by master builders who both designed and built the projects. As modern equipment and techniques developed, project schedules were compressed into several years rather than decades. As this process proceeded, designers were able to handle a number of projects concurrently, and the building process divided into design and construction.

Over the last 50 years, the design disciplines have further subdivided, and so separate disciplines, particularly in the engineering areas, function as separate operations under the general umbrella of design.

The design stage is divided into several discrete sequences. The first is known as the schematic, or sketch design, phase. As the names indicate, this is the preliminary definition, translating the program into definite physical parameters.

The next stage of the design, following approval of the schematic or sketch design, is the program or design development phase. In this stage, the sketch design is refined, and the basic layouts of space are brought into a semifinal situation. Following approval, the final development of design documents can begin.

The development of the design documents, or the working-drawing stage, requires the most time but includes the least number of decisions. Value analysis is best applied at the conclusion of the schematic and development stages, since change can be implemented at these milestone points with the least time and cost impact.

Designers are reluctant to make major changes following the working-drawing stage, since each change of even minor nature must be cross-coordinated with many disciplines. There is the opportunity for error as well as the added cost of change.

CONSTRUCTION

Most of the building projects in this country are constructed by contractors who have had no role in the design phase. Most contracts are awarded by sealed bid, negotiated, or on a cost-plus-fee basis. The specific method chosen may be mandated by legislative decree in governmental organizations or may be the result of prior experience in the case of companies. For instance, in the petrochemical field there is good precedent for the design-build approach, in which a complete project may be purchased. One term for this approach is "turnkey," which indicates that the contractor furnishes all design and construction required prior to turning of the key for the project.

Analysis of the conventional approach to strung-out, sequential stages of typical construction projects has led developers in the private sector to overlap projects. To a large degree, these developers depend upon the fact that they will build many projects over a period of time as a primary means of controlling their contractors. These contractors, in turn, are interested in repeat business at a fair price.

VALUE ANALYSIS: WHEN AND WHO

Obviously, the best point in the project cycle to apply a value analysis is the earliest point. However, for many reasons, this has not been the general point of application. In fact, the two major users of value analysis over the past 15 years have been the U.S. Army Corps of Engineers and the Naval Facilities Command. In both cases, the initial applications were through contractor value-engineering change proposals (VECPs).

While this area has the least potential, it does offer across the construction industry the potential for saving several billion dollars per year, certainly a worthwhile goal. In the case of the VECP, the contractor proposes and the construction manager and/or owner accept.

Where there are construction managers, value analysis should be organized and administered under their tutelage, but certainly not without the direct assistance of the design team and the ultimate approval of the owner.

In a multibuilding owner, such as the Veterans Administration, value analysis may be organized and administered under the owner's team rather than the designer, and in this instance the owner is usually acting as its own construction manager. The design team may also be an appropriate focus for the value-analysis effort.

6
BUDGET PHASE

In the great majority of cases, the players in the design and construction phases do not have a role in the budgeting stage. Nevertheless, this phase can offer the greatest opportunity for value analysis. Over many years, the development of budgets has been a selection process, generally practicing management by default; that is, roadblocks are placed in the way of certain undesired approaches, time is allowed to pass, and other delaying tactics are employed until the number of alternatives is reduced (on a de facto basis) to a manageable number.

In the early 1960s, the Department of Defense under Secretary McNamara developed the planning-programming-budgeting system (PPBS). The approach was developed in response to a definite need within the DOD, which was responsible for a multibillion dollar budget. The development of that budget was suffering from the results of over-organization. The very well-defined organizational breakdown structure within each service set the scene for many operational problems and even failures to perform. Budget requests were oriented to the requirements of a specific office and not to the operational needs of the various weapons systems or functional units. For instance, tactical air wings might not have nonnuclear weapons available as a result of cutbacks in a different service arm's requirement. Army airborne divisions might be battle ready without adequate airlift mobility. The obvious solution was an integration of requirements for the budget by utilizing a systemized planning approach. The Rand Corporation was given the assignment of developing this integrated approach to the statement of the results expected from a budgetary investment. The budget was no longer to be a static statement of demands, but rather a description of the functions and capabilities which the expenditure of money would provide. Should cuts be required, they would be applied sensibly to reduce a capability rather than reducing in a fragmentary way (often unintentionally) many elements common to many requirements.

The initial DOD PPBS efforts were so effective that in 1965 President Johnson mandated the use of PPBS in all executive budget development.

Unfortunately, PPBS was not quite ready for its own success. This system represented more of a concept than a technique. It was, however, an application on a conceptual basis of value analysis. Accordingly, the evaluation of the PPBS concept is directly related to the potential for value analysis in the budgeting phase.

From the can-do atmosphere and attitude of the Department of Defense, the transposition of PPBS into the other executive departments was often like planting the Biblical mustard seed in rock. Charles P. Schultz, director of the Bureau of the Budget under President Johnson, noted some of the problems faced by the nonmilitary agents in applying PPBS (value analysis) to the preparation of their budgets:

1. The problem of developing an experienced analytical staff appropriately aware of organized approaches

2. Difficulty in obtaining relevant data in format for utilization in performance-type budgeting

3. Problems involved in defining program benefits in specific enough terms to analyze on a value basis

The director of the budget indicated that the application of PPBS in the 36 agencies dealing with national problems resulted in great differences in technique and result. (Although his specific interest was PPBS, all forms of value analysis are subject to the same evaluation.)

The value analysis involved in PPBS had its beginnings in government, and most of the early experience was concerned with the military. While it is obvious that the Air Force cannot afford to have B-52s that run out of gas in midair because the operating line for flying tankers was not funded, it is just as true that civilian agencies need foresight and broadly based innovative planning. The electric refrigerator put a lot of ice plants out of business, and the retail coal yard is a thing of the past; these and other business applications could have benefited from a forward look.

Budgets can no longer afford to be merely a repeat of last year and probably never really could afford that luxury. Each budget item should be analyzed for value.

Capital budgets need intensive analysis, since costs of the design and construction are spread over many years and the tendency to invest in poor values is a constant temptation.

DEVELOPING ALTERNATIVES

The traditional planning process identifies needs but often fails to specify goals and objectives. Value analysis in the prebudget phase should go even further—it should develop specific alternative ways of meeting the goals and objectives so that the budget will reflect a specific response. This will result in better preprogramming specifics.

Budget Director Schultz posed a series of considerations which should be faced in developing a budget (Ref. 1). In the case of highways, he noted:

> We must ask primarily not how many of miles of concrete are laid, but more fundamentally what the program produces in terms of swifter, safer, less-congested travel—how many hours of travel time are eliminated, how many accidents are prevented. . . . We should be comparing the effectiveness of improvements to highways with that of additions or improvements to aviation and railroads as a means of providing safe and efficient transportation. This does not mean that we pick only one. . . . We do need to decide, at least roughly, which combination of alternatives is the preferred one. . . . In deciding to build an expressway through a downtown area, we must take into account not only the cost of the expressway, but also the cost of relocating the displaced residents, and, in a qualitative sense, the effects of the freeway on the areas through which it is to run.

As a result of value analysis, definition becomes progressively more specific. Taking the goals and objectives of the planning phase, the value team narrows down to the most viable alternatives and then further defines these, resulting in more availability of analytical information and therefore more opportunities for decision and further paring down of possibilities.

Henry Rowen (Ref. 2) reviews several diverse examples of the selection of alternatives:

> Our ability to support military forces abroad can be accomplished by the use of airlift, or sealift, or prestocking supplies at foreign bases, or through a combination of these techniques, which might differ from region to region or differ by type of commodity supplied. A new alternative might be invented, as it has, in the system which combines sealift with prestocking by having loaded ships kept on station abroad.

> The educational level may be raised by a variety of alternatives—increase in student participation, research and development in education, including the building of model facilities, improvements in the professional staff and in administration, the addition of new equipment, the reorganization of instruction, adult retraining, education in the home, education of the disadvantaged, and other categories of alternatives. Each of these categories, in turn, involves many alternatives. For example,

student participation can be increased through loan funds, work-study programs in industry, educational leads, starting public school at a younger age, new patterns of continued schooling aimed at the drop-outs, and many others.

The posing of reasonable alternatives requires creative thinking and the statement of many program possibilities which are easily dis-carded but nevertheless worthy of statement. Part of the PPBS art is the ability to state alternatives within the accepted planning goals. Since planning under the PPBS is not static, it is also likely that some restatement of policies and goals may be suggested and recycled for further consideration by the planners. This recycling must be carefully controlled, for if the programmers are to totally question the results of the planners, the result will be dynamic equilibrium instead of dynamic progress—in essence, going nowhere.

QUANTITATIVE MEASUREMENT

A simple quantitative method for analyzing the attractiveness of an in-vestment is establishment of the breakeven point: that is, the length of time over which the investment must be operated in order to recover all its costs. In formula fashion, this is stated as follows:

$$\frac{\text{Breakeven}}{\text{point (yr)}} = \frac{\text{investment}}{\text{cash benefit/yr}} = \frac{\text{investment}}{\text{annual income} - \text{annual cost}}$$

Example:

$$\text{Cost to set up program} = \$100,000$$
$$\text{Annual income} = \$200,000$$
$$\text{Annual operating costs} = \$150,000$$

$$\text{Breakeven point} = \frac{\$100,000}{\$200,000 - \$150,000} = 2 \text{ yr}$$

Assuming that all cash costs have been considered except taxes, the $50,000 apparent annual income would be reduced by about 50 percent because of corporate income taxes, with the following dramatic effect upon breakeven:

$$\text{Breakeven (posttax)} = \frac{\$100,000}{\$25,000} = 4 \text{ yr}$$

Breakeven point is usually a fictional evaluation used for comparative purposes only, since the inference of making money only after the breakeven point is overly simplistic. In actuality, particularly in a capital intensive investment, the cost of plant and equipment is spread out over many years, permitting the company or organization to recognize a greater income—and accordingly a greater tax liability— each year. For example, application of the simple breakeven formula for a capital intensive investment would be as follows:

Investment:

Set up of first costs	$ 50,000
Building	100,000
Machinery	200,000
Total	$350,000
Annual income	$250,000
Annual costs	$150,000

$$\text{Breakeven (pretax)} = \frac{\$350,000}{\$250,000 - 150,000} = 3^{1}/_2 \text{ yr}$$

This is less favorable than it should be because it recognized no salvage value for the building or equipment. Assuming that the breakeven point is actually closer to 2 years, and that the salvage or sale value of the building and machinery at that point might be as high as $250,000, the breakeven would then be:

$$\text{Breakeven (pretax)} = \frac{\$350,000 - \$200,000}{\$250,000 - \$150,000} = 1^{1}/_2 \text{ yr}$$

$$\text{Breakeven (posttax)} = 3 \text{ yr}$$

There is a substantial improvement in both pretax and posttax break-evens after insertion of the salvage value. This insertion, however, demands a substantial estimate in terms of the value of the property, and the characteristics of the product, the location of the plant, and the general nature of the economy would have tremendous impact upon the results of this calculation.

A more appropriate way for capital intensive investment to be shown is as recognition of depreciation:

$$\text{Breakeven} = \frac{\text{fixed cost} + \text{annual cost of capital investment}}{\text{gross annual cost}}$$
$$- \text{ annual operating cost, including depreciation}$$

where fixed cost = $50,000 (one-time set up)
  building = $100,000 (straight-line depreciation; no salvage value; 20-year life costs, $5,000/yr)

equipment = $200,000 (straight-line depreciation; no salvage value; 10-year life costs, $20,000/yr)

x = breakeven years

Breakeven: $x = \dfrac{\$50,000 + x(\$5,000 + \$20,000)}{\$250,000 - 150,000 - 25,000} = \dfrac{\$50,000 + 25,000x}{\$75,000}$

x = 1 yr (pretax)

Now assume that there is an opportunity to lease the equipment and building at a cost annually of $50,000 and that the only investment need be the fixed cost of $50,000. The breakeven point would then be:

$$\text{Breakeven (pretax)} = \dfrac{\$50,000}{\$250,000 - \$200,000} = 1 \text{ yr}$$

The reverse of these breakeven points is greater return, which can be considered reciprocal of breakeven, and so the rate of return for a 4-year breakeven would be 25 percent, while that for a 1-year breakeven would be 100 percent.

DISCOUNT RATE

Rate of return on investment is a major evaluation factor. Where the investment involves borrowed capital, the interest on that capital is properly a part of the cost and can be accommodated in that fashion. However, where the selection of a program or project is using internal resources, which may include working capital, failure to include recognition of this use can lead to the selection of the wrong projects.

The value of the use of internal resources, whether governmental or private industry, is called the discount rate. This amount should be discounted from the return on investment, or depending upon the method of calculation, can be added to the cost of implementing the program.

Private industry usually uses a rate on the order of 15 percent discount, pretax, as follows:

Sector	Assets (trillions, $)	Annual rate of return, %
Manufacturing	275	17.7
Utilities	94	9.4
Transportation	34	6.0

Note the lower rate of return in the highly regulated industries. Schultz (Ref. 1) estimated that a majority of 53 Corps of Engineers and Bureau of Reclamation projects which he had examined would have been rejected had a discount return rate of 10 percent been required, even though these organizations pride themselves on the return from the major water and power programs which they undertake. Tax income to all government sectors is reduced when a public project replaces one which could be accomplished in the private sector.

SUMMARY

In this section, the terms PPBS and value analysis have been utilized interchangeably. Actually, most budgeting is devoid of both processes. PPBS is a concept within which value analysis can function effectively in prebudgeting considerations.

The development of a sound budget combines needs (goals and objectives), cost, scheduling, and function. As the activity becomes more organized, it will naturally follow that design professionals and builders will have more of an opportunity before the major project parameters are established in the form of a budget which all parties must live with.

There are certainly two sides to this page. On one side, the policymakers sometimes are concluding that brick and mortar facilities are really not needed. This decision may be occasioned by major breakthroughs, such as mental health medication, which has tremendously reduced the physical holding capacity previously required in mental hospitals.

In another specific situation, the state of New Jersey was about to enter into a crash building program to house more prisoners, but on reflection the director of the state prison system and his advisors revamped the criteria for prisoners to be held in pretrial, confinement, bail, and probation, thus drastically revising the holding capacity required in the state over the short run. This relaxed what would have been a major crash design and building effort which could have produced a poor investment in state funds.

There is another side to the page, and that side suggests that the policymakers would be wise to bring in design and construction professionals in the prebudget stage to provide realistic information in terms of what can be done in terms of time and cost so that the budget parameters are realistic ones which can be implemented.

The case study which follows is an example of prebudgetary value analysis. The analysis led to an unexpected answer for a serious urban problem.

REFERENCES

1. Schultz, Charles P.: *Hearings before the Subcommittee on National Security and International Operations of the Committee on Government Operations,* U.S. Senate, Aug. 23, 1967.

2. Rowen, Henry: *Hearings before the Subcommittee on National Security and International Operations of the Committee on Government Operations,* U.S. Senate, Aug. 23, 1967.

CASE STUDY C
BUDGET PHASE (RAT CONTROL)

ENVIRONMENTAL CONTROL—RATS

A study team was assigned by the author to the task of developing a program to define a budget line to be appropriated for better rat control in urban areas.

The initial analysis was based upon the verb-noun functional description "Kill Rats."

A study team was established which included systems analysts, engineers, and a scientific investigator with a background in ecological and biological studies. The team was instructed to emphasize an ecological approach.

In reviewing the basic functional premise, ecologists pointed out that a program of poisoning or killing by other means would undoubtedly be unsuccessful, even though it represents the typical approach over many decades.

One reason for the tendency to utilize extermination as the implementation of "Kill Rats" is its directness and visibility to the public.

Substantial improvements have been made in chemical poisons which kill on a basis which seems to mask the ability of the rat population to discern the poison. One such poison is Warfarin. However, even with advanced poisons, measurements of urban rat populations reveal the inefficiency of all methods of poisoning and trapping. One of the first measurements which determined this was an effort in Manila headed by the U.S. Public Health Agency which failed to significantly reduce rat populations or to impede the course of rat-carried diseases, despite the fact that a host of cheap labor was made available. That same measurement did point out that some level of success was achieved when plague-carrying rodents were systematically pinpointed, tracked, and killed indirectly by reduction of the rat access to shelter and food.

Many biological sources confirmed that in a given environmental situation, at a given time, the ability of a habitat to support a particular species population is finite. Therefore, there is a point beyond which the rat population will not increase. This ecological concept—the carrying capacity of an environment—is the cornerstone of modern wildlife management. It recognizes that ecologically there is an equilibrium between the habitat and the species.

Recognizing that the rat population would become self-limiting because of environmental factors, the ecologists shifted the function of the program to "Control Environment." The focus of the study now considered the control of the rat population from an entirely different viewpoint—altering their habitat. Permanent control can be achieved through an environmental approach which eliminates the rats' environmental necessities: food and shelter. Federal guidelines for rat-control project grants under PL89-749 give priority to the environmental approach: "Project activities should include more than just rat extermination activities, . . ." The federal approach indicates that the scope of a well-designed rat program should include most or all of the following elements:

Communitywide plans

Community organization and citizen participation

Community education

Effective administrative organization and interagency coordination

Code enforcement

Evaluation

Because of the recognition of the importance of harborage and food supply, the planning factors for the rat-control program became much broader, and the realization that narrow programs were a waste of money and resources led to short-range savings but presented the broader problem of control through improvement of the physical environment.

The planning for rat control was oriented specifically to New York City, and the two areas of food supply and harborage were considered separately. Liaison with the city's sanitation and solid-waste disposal programs was established, and it was agreed that the existing programs of this department would be focused in any area of treatment to achieve the desired results in terms of control of food supply. The study then focused on the question of harborage, in particular abandoned buildings. The problem was not what to do but where to start. The analysis considered a test program to view 800 blocks as a sample, with inspectors visiting each block and identifying each abandoned building. From the collected data, abandoned buildings would be identified and then located. The program would then make managerial decisions in regard to abandoned buildings—specifically, whether to demolish or preserve.

The individual building decision affecting thousands of buildings tended to be subjective—with one camp suggesting clear and pave (to secure from rubbish and prevent brickbats), while others indicated maintaining the buildings by sealing off doors and windows (against the time when they could be rehabilitated). The economics were not necessarily clear, and a decision tree was established for a building-by-building evaluation. This is shown in Fig. C-1 and starts with the question "Should we demolish this building?"

It then goes on to describe 34 ways to attack the problem of an abandoned building. Each path produces a different result, with the result of any given

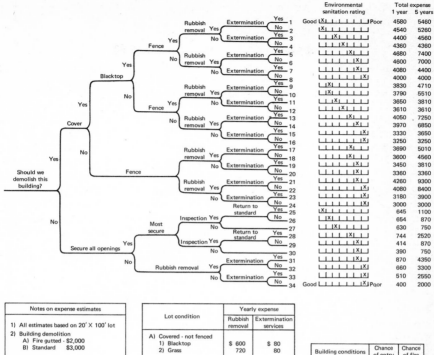

FIGURE C-1

Decision tree established for urban renewal (from a report for the city of New York, James J. O'Brien, P.E., in *Management Information Systems,* Van Nostrand, New York, 1970.)

Notes on expense estimates
1) All estimates based on 20' X 100' lot
2) Building demolition
A) Fire gutted - $2,000
B) Standard $3,000
3) Cover lot
A) Blacktop $1,000
B) Grass $ 250
4) Fence lot
(7' high - 1' barbed wire) - $360
5) Secure openings in abandoned buildings
A) Most secure (brick) - $600
B) Secure (plywood and tin) - $300

Lot condition	Yearly expense	
	Rubbish removal	Extermination services
A) Covered - not fenced		
1) Blacktop	$ 600	$ 80
2) Grass	720	80
B) Covered - fenced	180	40
C) Uncovered - not fenced	1080	180
D) Uncovered - fenced	240	90
E) Abandoned building		
1. Unsecured	480	240
2. Loosely secured	120	60
3. Tightly secured	000	000

Building conditions	Chance of entry	Chance of fire
A) Unsecured	90%	20%
B) Loosely secured	25%	10%
C) Tightly secured	5%	1%

path subjectively expressed on a 7-part "good" to "poor" skill. Any expense required to complete the actions on the path are shown for 1 year and 5 years, respectively. Since resources in all such programs are limited, the purpose of the decision tree is to develop information which leads management to choose the best approach in terms of cost effectiveness. The decision tree indicates, for instance, that path 25 will permit similar "good" results for five abandoned buildings for a $5,500 investment, while path 1 permits only one good result for one abandoned structure for the same $5,500 investment. The operational decision tree becomes more complex and must take into consideration the basic integrity of the structure. However, most of the abandoned structures can be secured by being left standing but protected against intrusion.

Field inspections in the ghetto areas revealed that paths 24 and 34 are most heavily relied upon currently. Path 24 spends money to solve the problem, but little impact is made because a refuse-strewn vacant lot is just about as bad environmentally as an abandoned building. Path 34 is characterized as "do nothing" but still involves expenditures.

The assembly of information on all the houses in a block results in a score by block and may reverse the individual building decision in favor of total demolition—for instance, in the case where only two or three buildings in a block are marked for preservation. The budgeting decision still remains for a return on investment analysis of resources available versus problems to be solved. The extreme views include, first, that which would put all resources in the poorest score in order to hold the line; the second approach puts the same resources into more houses in better shape to achieve maximum initial results. The value approach views across the entire problem, compares resources, and suggests decisions which will produce the most lasting results in the context of the policies and objectives established.

CASE STUDY D
BUDGET PHASE (UNIVERSITY HOSPITAL)

In 1965, the trustees of an Eastern university decided to sponsor a major construction program to modernize the university hospital. In addition to the delivery of health care, the university hospital provided a focus for research projects and worked in close cooperation with the university medical school. In deciding to go forward, the trustees specified a number of objectives including:

More efficient delivery of health care

Retirement of obsolete facilities

Retirement of nonfireproof facilities

Improved parking facilities

A well-qualified hospital consultant was retained, and a program developed in less than 1 year. After acceptance of the program, the trustees awarded a design commission to one of the leading architects in the Northeast for design of the centennial program.

Schematic design which would ordinarily have taken no more than 6 months dragged into several years, as a 17-member doctors committee worked directly with the designer to rearrange the schematic to meet both the program and the requirements of the committee. One problem was a continual change in membership of the doctors committee, with a corresponding delay in progress.

The initial budget goal was $25 million for the program. As the various requirements were introduced into the schematic design phase, over an extended period of time, the overall figure grew to $105 million estimated construction cost. At this point, the trustees insisted that the cost be dramatically reduced, and a revised plan was developed and costed out at $63 million.

This development period had utilized 4 years, and the medical committee was diligently continuing to work; the schematic design phase was very slow in coming to a complete definition. A substantial amount of redesign was required, and the designer was compensated on a cost-plus basis, which was the only equitable method of handling the continuing costs of the schematic phase.

The author was brought into the project as construction manager, and a number of utility relocations were identified as long-lead items suitable for fast tracking. Investigation indicated that the city would be unable to design and implement the utility relocations in less than 2 or 3 years. Utilizing a unique financing approach, it was possible to initiate these programs within 6 months and to complete major relocations within an additional year.

In the meanwhile, however, the construction manager conducted a value analysis of the overall program.

The principal functions of the plan were:

Health care

Support education

Support research

In the delivery of health care, secondary goals were: practice of preventive medicine, delivery of community care, and support of private medical practices. These functions were being carried out with the existing plant. A major goal of the new program was "improve environment."

In support of the medical school, one of the major functions was, again, "improve environment." The environment was to be improved to enhance learning and to increase staff morale. Also, the improved medical school would draw, through the university hospital, more interesting cases which are appropriate to a major medical school.

The "support research" function was oriented toward attraction of interesting projects and independent funding for operating costs. These in turn would attract more projects and research staff.

In the administrative area, a joint integrated medical facility for servicing at least two hospitals had been included in the program. This shared facility would provide better food services, storage space, and maintenance capabilities.

HEALTH CARE FACTORS

Category	Factors	Opportunities to increase revenue	
		Yes	No
1. Health delivery	1.1 Preventative medicine	X	
	1.2 Community care	X	
	1.3 Private practice	X	
	1.4 Improved environment	X	
2. Medical school	2.1 Interesting case		X
	2.2 Attractive to staff		X
	2.3 Environment for learning	X	
3. Research	3.1 Interesting projects	X	
	3.2 Independent funding	X	
	3.3 Attract projects	X	
	3.4 Attract staff		X
4. Administration	4.1 JIM { food	X	
	stores	X	
	maintenance	X	
		11	3

FIGURE D-1
The four major categories of construction.

DECISION TREE

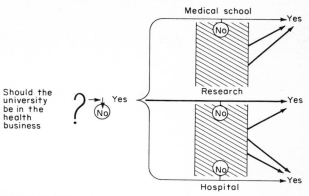

FIGURE D-2
Decision tree.

Figure D-1 summarizes the four major categories of construction and the secondary factors relating to each one. Eleven out of fourteen had potential opportunities for increase in revenue. However, a detailed evaluation indicated that the substantial cost reduction had reduced the capability to recover increased revenues sufficient to offset the new increased debt service. The inevitable result was to be an increased deficit operation for the university hospital.

In continuing the value analysis, the construction manager chose to start at the highest level of the value hierarchy. The basic function addressed was delivery of health care. The question addressed was whether the university should be in the "health business" (Fig. D-2). If the answer was no, certain problems were immediately posed, since programs were in progress (Fig. D-3). In terms of the hospital plant, it would have to be demolished, con-

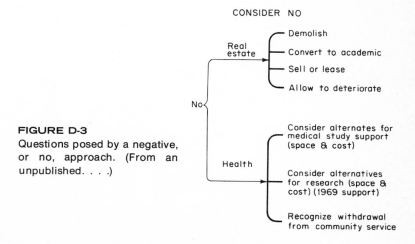

FIGURE D-3
Questions posed by a negative, or no, approach. (From an unpublished. . . .)

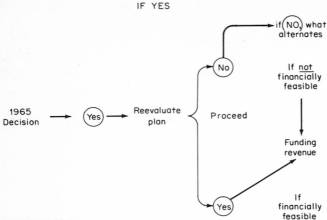

FIGURE D-4
Chain of events possible with a positive, or yes, approach.

verted to other uses, sold or leased, or allowed to deteriorate. Each of these had difficult connotations, and all would ultimately cost the university money.

In the area of health, a decision to retreat from the health care required alternatives for medical study, support for the medical school, alternate approaches for research, and recognition that the university would be withdrawn from community service—an obligation of service which it had taken

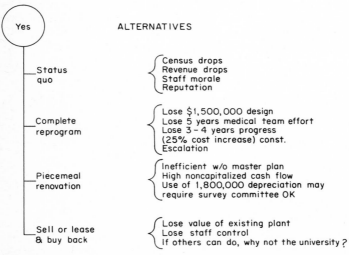

FIGURE D-5
Possible alternatives.

on as part of a broadening awareness of the social responsibilities of leading institutions.

The ramifications of a negative approach having been evaluated, Fig. D-4 illustrates a chain of events possible with a positive, or yes, approach. The basic 1965 decision could be reevaluated and implemented if financially feasible. If not feasible, Fig. D-5 shows the following possible alternatives:

1. Maintain Status Quo Patient census would drop, revenue would drop, staff morale would be affected, and the reputation of both the hospital and the university would suffer. Obviously, this was not the approach to use.

2. Completely Reprogram This would result in the loss of 1.5 million in design effort, 5 years of a combined team effort, and 3 to 4 years of progress and would require any ensuing construction to reflect a 25 percent cost increase because of escalation.

3. Piecemeal Renovation This approach would show some improvement. However, it would be inefficient without a master plan (which did not exist at this point in time), would require high noncapitalized cash flow (assuming that renovations occurred and were funded through operating monies), and would require approval of the regional health-care agency.

4. Sell, Lease, Buy Back The possibility of selling the plant to a private firm and leasing it back would lose control of the staff and virtually give away the value of the existing plant; and if others could properly manage, why not the university hospital.

These basic value questions were presented to the trustees with the recommendation that the present plan be written off and that an interim plan of development be established rapidly. A number of basically correct construction programs, such as the utilities relocations, were continued, and over a number of years the objectives are gradually implemented. The effect has been to maintain service, develop the capital plant gradually, and approach the value of the original plan at a small fraction of the original cost.

The analysis performed would actually have been more timely at either the budgeting or programming stage. However, even though an extreme example, it does indicate that value analysis can point out the appropriate action to be taken even at a relatively advanced stage of commitment.

7
DEVELOPMENT

In development, the cost of land, tax rates, and the cost of money are functions of the time and situation. The costs which can be controlled, and should therefore be included in the value analysis, are the initial cost of construction, the cost of operating, and the return.

Value analysis—delivering the best value at lowest cost, while keeping the life-cycle costing in the context of the prime function, which is *optimum return*—is within the general parameters of value analysis for projects. The analysis of potential returns is an area which may be neglected. Developers are susceptible to intuitive determination of project mix. From the marketing viewpoint, they discern that a certain type of combination of facilities will present a picture to the potential buyer or renter of the development. The syndrome of the marketing image is analogous to aesthetics and the design team. It has a definite value, but that value should be analyzed in the context of the overall complex formulas which affect the bottom line.

The marketing image can adversely affect the life-cycle costs of a development. In condominium complexes, developers striving for an interesting marketing image include clubhouses and other amenities which must be maintained by the purchasers.

VALUE ANALYSIS

Value analysis has a different meaning to the developer. A project simply will not proceed if evaluated on the same total-systems basis to which a governmental project must react; that is, if costs are equated to the total life utilization, the 92 percent figure developed by the GSA would no longer be represented by salaries being paid out but would represent rentals paid by occupants. The developer will probably be with the building for less than 25 percent of its life span (after depreciation has run out) so that the maintenance costs and operating costs paid by the developer would be on the order of 1 to 2 percent of the overall total life-span cost of the building.

Accordingly, to the developer value is a negotiable item. The developer is willing to balance lower initial cost against higher operating costs, seeking an optimum balance between profit and the marketability of the units. The interplay of factors is very complicated. There is no single correct answer, since the solution is a heuristic one. Computerized cost models have proven very useful in development of ranges of solutions. This type of responsive model is necessary, since the factors, particularly the cost of money, can vary from day to day.

CASE STUDY E
DEVELOPMENT FEASIBILITY ANALYSIS (FLORIDA)

DEVELOPMENT FEASIBILITY ANALYSIS

The following feasibility analysis is an actual one prepared to evaluate the potential for developing a 9.1-acre property in central Florida.

The total site area of 396,300 sq ft (9.1 acres × 43,560 sq ft/acre) was to be developed as shown in Table E-1.

After the basic ground coverage of the project had been laid out and it had been ascertained that the coverage for the site was reasonable in terms of local building codes, an estimate of costs was prepared (see Table E-2).

Income calculations for the facility were developed as follows.

Occupancy rate	70%	75%	80%	85%
Room sales	$1,277,500	$1,368,750	$1,460,000	$1,551,250
Restaurant net	204,400	219,000	233,600	248,200
Miscellaneous	38,325	41,063	43,800	46,538
Total	$1,520,225	$1,628,813	$1,737,400	$1,845,988
House profit	684,111 (45%)	749,725 (46%)	816,578 (47%)	886,074 (48%)

Table E-1

Facility	Area, sq ft
Existing bank	10,000
Office building	18,000
Theater	5,500
Motor inn	10,000
Apartment	12,000
Retail department store	75,000
Retail specialty shops	40,000
Parking (two levels)	135,000
Site development	90,800
Total	396,300

Table E-2

Facility, gross sq ft	Unit cost, sq ft	
	Low	Moderate
Direct costs		
Office tower	$24	$32
483,900	11,613,600	15,484,800
Theater	20	25
5,500	110,000	137,500
Motor inn	28	34
100,000	2,800,000	3,400,000
Apartments	17	21
140,000	2,380,000	2,940,000
Retail department store	14	18
150,000	2,100,000	2,700,000
Retail specialty shops	15	19
40,000	600,000	760,000
Parking (850 spaces)	7.30	9.50
270,300	1,973,200	2,597,900
Site development	6	10.50
(plazas, walkways, landscaping)		
100,000	600,000	1,050,000
Subtotal direct costs	22,176,600	29,070,200
Construction contingency (4%)	887,100	1,162,800
Total direct costs	23,063,700	30,233,000
Indirect costs (percentages are multiplied by 1.3 above)		
Architecture and engineering (4%)	922,500	1,208,000
Program management (4%)	922,500	1,308,000
Taxes during construction (3 yr)	1,388,000	1,810,000
Insurance during construction	20,000	28,000
Interim financing (34,000,000 or 43,000 × 3 yr × 7% as used)	3,570,000	4,640,000
Finder's fee	250,000	250,000
Permanent loan fees	450,000	569,000
Legal costs	50,000	50,000
Closing costs	20,000	25,000
Completion bond	200,000	262,000
Title insurance	30,000	30,000
Appraisals	25,000	33,000
Leasing commissions	450,000	450,000
Initial losses and leasing discounts	400,000	500,000
Accounting fees	25,000	25,000
Subtotal indirect costs	8,723,000	11,238,000
Project contingency (1%)	87,200	112,400
Total indirect costs	8,810,200	11,350,400
Land cost	2,340,000	2,340,000
Total project cost	$34,116,900	$43,923,400

1. Hotel

200 rooms, average room rate $25 (high double occupancy). House profit esti-
mated at 45 to 48 percent of the sum of room sales plus restaurant net plus
miscellaneous. Restaurant sales estimated at 80 percent at room sales. Restaurant net estimated at 20 percent of restaurant sales. Miscellaneous sales estimated at 3 percent of room sales.

Real estate taxes @ $0.75/gross sq ft × 100,000 gross sq ft, would reduce
house profit by about $75,000. Summary income based on 75 percent occu-
pancy. Income is $749,725 less $75,000, or $674,725.

2. Office Tower

Rental rate is based on $6.50 per net leasable area and is a net rate. Occupants
would pay in addition the sum of operating expenses plus real estate taxes, the
total of which would approach $2.25 to $2.75/sq ft leasable area.

Total leasable area = 383,000 sq ft

Rental at full occupancy (383,000 sq ft × $6.50/sq ft) ×	$2,489,000
Less 5% vacancy	124,800
Net income	$2,364,700
Note:	
Operating costs estimated at 383,000 × $1.75	= $ 670,000
Real estate taxes at $0.75/gross sq ft × 483,900 sq ft	= 363,000
Total	$1,033,000

$$\text{Effective rental} = 6.50 + \frac{1,033,000}{383,000 \times 0.95} = \$9.35/\text{sq ft (leased)}$$

3. Motion Picture Theater

Estimated 25 percent use rate, ticket price of $2.50, a 25 percent profit margin
(including refreshments), and six screenings per day.

Net income = 365 days × 6 screenings/day × 600 seats × $2.50/ticket × 25
percent use factor × 25 percent profit margin = $205,300
Real estate taxes = $5,500 × $0.75 = $4,100

Note:
Entertainment tax of approximately $20,000 also included in total.

4. Apartments

Income:

50 Efficiencies @ $220/mo × 12 mo	= $132,000
100 1-Bedroom @ $300/mo × 12 mo	= 360,000
50 2-Bedroom @ $395/mo × 12 mo	= 242,000
Total income potential	$734,000
Less 5% vacancy factors	36,700
Income	$697,300

Expenses:

Operating administrative @ $1.50/sq ft × 140,000 sq ft = $210,000

Real estate taxes @ $0.75/6 sq ft × 140,000 sq ft   =   105,000

Total expenses $315,000

Net income before debt service $382,300

5. Retail—Department Store

Assumed proprietory lease @ $4/sq ft plus operating costs leased
to one tenant (150,000 gross sq ft @ $4/sq ft)   = $600,000

Less real estate taxes @ $0.75/sq ft   =   112,500

Net income before debt service $487,500

6. Retail—Specialty Shops

30,000 net @ $7/sq ft plus operating costs (lease × 95%
occupancy)   = $200,000

Less real estate taxes @ $0.75 × 40,000   =   30,000

Net income before debt service $170,000

7. Parking

Requirements	Day	Evening
Office building and bank	375	25
Theater	25	150
Motor inn	50	120
Residential	50	200
Retail—department store	300	300
Retail—specialty shops	40	40
Total	840	735

Build for maximum daytime demand, or 850 spaces

Transient income = $1.25/space/turnover

Monthly income = $35/space leased

Parking Income:

Office:

50 spaces × 12 mo × $35/mo   = $ 21,000

32 spaces × 220 days × 1$\frac{1}{2}$ turns × $1.25/space   =   134,100

Theater:

150 spaces × 365 days × 2 turns × $1.25/space   =   136,900

Motor inn (included in hotel analysis):

120 spaces × 365 days × $1.25/space   =   54,800

Apartments:

200 spaces × 12 mo × $35/mo   =   84,000

Combined retail:

340 spaces × 312 days × 1¹/₂ turns × $1.25/space =   198,900

Total garage income $629,700

Expenses:

850 spaces × $110/space   = $ 93,500

Real estate taxes @ 850 spaces × 318 sq ft/space × $0.75/sq ft =   202,800

Total expenses   296,300

Net income before debt service $333,400

8. Summary—Income Calculations

	Gross Income	Operating Costs	Net Profit (before debt service)
Hotel	$1,628,813	$ 945,088	$ 674,725
Office tower	3,397,700	1,033,000	2,364,700
Theater	821,200	615,900	205,300
Apartments	697,300	315,000	382,300
Department store	600,000	112,500	487,500
Specialty shops	200,000	30,000	170,000
Parking	629,700	296,300	333,400
	$7,974,713	$3,357,288	$4,617,925

9. Financial Calculations

	Low	Moderate
Total project costs (Exhibit 3)	$34,116,900	$43,923,400
Net income before debt service (Exhibit 4)	4,617,100	$4,617,100
Mortgage @ 70% financing	$23,881,800	$30,746,400
Equity required @ 70% financing	$10,235,100	$13,177,000
Debt service (0.0950) @ 70% financing	$2,268,800	$2,920,900
Debt service (0.0950) @ 100% financing	$3,241,100	$4,172,700
Net income after debt service:		
70% financing	$2,328,300	$1,696,200
100% financing	$1,376,000	$434,400
Return on equity (item 7/item 4):		
70% financing	21.8%	12.9%

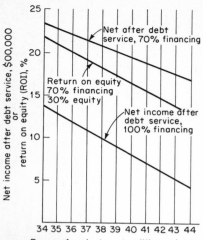

Net income after debt service, $00,000
or
return on equity (ROI), %

Net after debt
service, 70% financing

Return on equity
70% financing
30% equity

Net income after
debt service,
100% financing

34 35 36 37 38 39 40 41 42 43 44
Range of project cost, millions of dollars

FIGURE E-1

Income vs. project cost. Note that "Return on Equity" calculation does not consider the impact of taxes (negative) which might tend to increase the return. Also, it does not consider the time value of money or the expected measure in real property values over the holding period. This kind of analysis would be necessary for internal purposes of the developer but not for financing.

PROGRAMMING PHASE

Programming is the phase of the construction project which follows the budgeting and precedes the commencement of actual design work. The importance of this category is often unrealized, and in many cases programming may be omitted. Unfortunately, these are usually cases in which the budgeting effort has been performed on an intuitive basis rather than through a systematic approach such as PPBS. The program provides a factual framework of requirements within which the design is to be performed. It describes, for the owner, the requirements which the facility is to meet. It may also describe the general thrust in which these requirements are to be met. Programming is often subdivided into two stages: functional requirements of the owner and an architectural program within which the design is to be performed.

Where PPBS has been utilized in developing the budget, and the project is one with a relative certainty of receiving budget approval, programming may precede the formal acceptance of the budget. This is particularly useful in governmental situations in which budget approval may require 6 to 9 months. In effect, the early start of programming is a fast tracking of the predesign stage. The risk in money is minimal, and the time recovered is purchased at a very low cost.

The need for programming is recognized by the experienced owner. Organizations which utilize programming include the Veterans Administration, General Services Administration, the Department of Health, Education and Welfare, and the Department of Defense as well as cities such as New York.

Inexperienced owners, either through optimism or failure to understand the total budget-design-build process, may omit the programming phase. The resultant cost in time and money is inevitably a high one. Designers presented with a project in an unprogrammed state have learned through the years that it is wise to proceed cautiously. Since the owner has not actually defined his requirements in complete enough detail, the designer must undertake that task. The owner may appear to have gained by transposing some of his responsibilities to the designer with no increase in fee. However, he pays in a dragged-out design phase. The designer is forced into this position. To do otherwise will require that he concurrently define the scope of the work to be accomplished and initiate design. In almost every case, this results in a need to change the design. Obviously, the most economical approach for the design is to hold back the start of the actual design work while developing the program. Since he is not being compensated at appropriate design fees for programmatic development, his input is on a very limited basis.

Programming does require effort, and effort costs money, whether internal or through consulting. However, the investment should be a very worthwhile one. The programming phase is perhaps one of the greatest opportunities for value analysis, but the owner should be aware that he has to specify the value work if he expects it included in the program phase, particularly if the work is done by a consulting organization.

The hospital field is one in which development of programs is well recognized. Much of this is as a result of the evolution of hospital consulting. This field of consulting has been encouraged by the HEW management. The approach is a good one, since it provides the federal government, supporter of many hospital construction programs, with a detailed explanation of the need for a facility and with a clear functional program which describes the manner in which that need can be met by this specific project. The cost for this combined statement of need and program to meet the requirements usually falls in the range of 1 percent of the projected completion cost. Typically, the programmatic work by the hospital consultant does not include value analysis. The owner would have to specify this as an additional scope of work, either by his staff or a consultant. HEW recognizes the value-analysis approach and would encourage the use of value analysis at this point.

An important inclusion in programs is any appropriate statement of owner policy. For instance, the Environmental Protection Agency (EPA) is including its document *Guidance for Facilities Planning* as part of its grant background material. By endorsement, the guidelines are also imposed upon the design team. James W. Meek, acting chief of EPA's Planning and Standards Branch, was quoted by *Engineering News-Record* as saying:

In light of the recent $60-billion needs survey, current inflation, the relatively limited funds available for treatment works, and the shortages of energy and materials, it is imperative that all concerned interests work together to maximize environmental gains from pollution abatement investments.

Value analysis by the EPA has led to post-budgeting changes which have resulted in substantial savings. *Engineering News-Record* (March 14, 1974) cited several examples in which cost effectiveness reviews reviews have led to savings:

1. In Glendale, Colorado, the proposed treatment plant construction plans were changed, resulting in a cutback in costs from $1,500,000 to $200,000. The city had intended to build a new plant, but the EPA regional office suggested that additional sewage load could be hooked on to a nearby Denver intercepter. That agreement was made, and Glendale had only to upgrade treatment for its existing facilities.

2. In Aliso, California, EPA suggested that the population projections were too high and that the project should be evaluated on a lower population. The result was a reduction in size of two critical interceptors, one outfall and one land outfall. The reduction in scope saved $5 million.

3. In East St. Louis, Illinois, a proposed plan to upgrade five plants for five communities was changed so that four communities were consolidated into one plant. This was a functional value analysis and resulted in cost savings of almost $7,000,000.

4. In New Haven, Connecticut, a plan to upgrade a plant from primary to secondary treatment was changed because the additional construction required filling in of wetlands. Flow was switched to another plant. That plant was enlarged, but the shift in planning saved the city more than 13 percent of the cost.

The General Services Administration has pioneered in the area of life-cycle savings. Buildings are evaluated on their total life cost rather than simply on the initial costs. This value-oriented approach makes obvious sense for well-capitalized owners. All owners are well advised to consider the approach, but in the private sector compliance would not be mandated, unless it is affected by the federal energy standards.

Certainly, the program should include a statement in regard to the life-cycle savings objectives to be observed in the design. Energy standards have become more recognized for their importance with the higher cost of energy sources. The GSA has issued *Energy and Conservation Design Guidelines for Office Buildings,* which includes a parameter of 55000 Btu/gross sq ft/yr. Inclusion of this as a policy document would be a substantial guideline in the program. Another document prepared by the National Bureau of Standards (NBS) called *Draft De-*

sign and Evaluation Criteria for Energy Conservation and New Buildings has been in preparation for some time, with the final version being reviewed and revised by the American Society of Heating, Refrigerating and Air Conditioning Engineers (ASHRAE).

The Tishman Research Corporation conducted a survey of 86 office buildings in New York City, indicating a median consumption of 160000 Btu/sq ft/yr. Tishman Research feels that the short-term budget limit for energy use should be in the 100000 to 120000 Btu/yr range. Libby-Owens-Ford Glass Company sees the 55000 Btu as a good long-range objective. GSA Administrator Sampson indicates that the 55000 Btu/sq ft/yr is a flexible goal.

FUNCTIONAL PROGRAM

The functional program is that portion of the program which specifies the spatial distribution required and may specify the relative locations. The National Bureau of Standards has developed a series of formats to be used in identifying the building elements and parameters. These are shown in Fig. 8-1.

In developing this documentation, the NBS staff surveyed a number of buildings, indicating that for a 40-year life the total life cost of a building is:

Cost of people using the building (salaries)	92%
Maintenance and operation of the building	6%
Initial building cost	2%

The point is that life-cycle costing is impacting the 6 percent and affecting directly the efficiency of the utilization as reflected in the 92 percent.

Owners who build many projects often reflectively develop parameters for future projects. Obviously, this is an excellent position for the application of value analysis, since it will affect many projects downstream. Conversely, there is concern that criteria may become inflexible, precluding the opportunity to apply value analysis.

The Educational Facilities Laboratories funded by the Ford Foundation have done much to develop this type of reflective parameter. With their help, the school district of Philadelphia acquired the services of Stanford Research for the development of standard programming factors in their several hundred million dollar school building program over the past decade. These universal functional programs are adjusted to suit the individual project. Also in Philadelphia, in 1971 to

Top matrix (building process / support systems / built elements):

Building process				
Qualifying				
Manufacturer	*The process*			
Shipping				
Construction				

Support systems				Built elements: hardware						
Life	Task	Psych.		Structure	HVAC	Utilities	Finished Floor	Luminaires	Finished Ceiling	Space Dividers

Required spatial attributes:

			Conditioned air							
			Illumination							
			Acoustics							
	The user		Stability	*Building "in use"*						
			Health and safety							
			Planning							
			Task support							
			Esthetics							
			Maintenance							
			Interface							

Built elements

	A Structure	B HVAC	C Utilities	D Finished Floor	Luminaires	F Ceiling	G Space Dividers
1 Conditioned air							
2 Illumination							
3 Acoustics							
4 Stability and strength of materials							
5 Health and safety			"In use" description				
6 Planning							
7 Activity support							
8 Esthetic							
9 Maintenance and improvement							
10 Interface							

Special attributes required by the user

FIGURE 8-1
Formats for identifying building elements and parameters. (From a National Bureau of Standards study for PBS/GSA, used in *Contractor's Management Handbook,* McGraw-Hill, New York, 1971.)

1974, the Hospital Survey Committee, regional reviewer for HEW, commissioned a design team to develop factors for use in the preparation of functional programs. (This study is described in detail in the Case Study F at the end of this chapter.)

In scheduling school classrooms, computers have successfully been used to most effectively utilize the classroom space available. The reverse of this can be implemented to preproject the number of classrooms needed. For example, Fig. 8-2 shows the calculations for an elementary school of 480 students, with 40 sessions, or periods, per week. In this example, music and art are held one session per week per class, and there are 12 classroom groups. The auditorium is used on an as-required basis, and each classroom group has gymnasium twice per week. The cafeteria is used on a daily basis by class.

This illustration demonstrates the relative inflexibility of the smaller school unit. Functional programming would demonstrate that individual auditoriums and gymnasiums are much more easily justified for larger groupings and that individual room utilizations become more effective. In larger schools, this type of study can be used to demonstrate that cafeterias cannot adequately double as auditoriums or gymnasiums.

The functional program should transmit the locational requirements of the owner. Figure 8-3 is an adjacency matrix. This was developed for an actual school administration building. A low score of 0 was given if a definite relationship was not required on a regular basis. A relationship requiring contact on about a weekly basis was given a score of 1. A mandatory adjacency relationship was scored 2. Of course, a spread in scoring of 1 to 10 or any other scale can be utilized. In Fig. 8-3, a high score indicates a need for proximity. The matrix is plotted showing individual functional departments with scoring relative to other departments.

In the portion of the matrix where the department is double described, an X is placed to indicate the nonapplicability of the space. The matrix will be a triangular one and can be prepared in triangular form with half of the square or matrix omitted.

The heavy line in Fig. 8-3 shows the relationships for the controller-treasurer's office. Nonrelated areas include curriculum-oriented and academic areas. The overall rating for this office is determined by adding horizontally to point X and then vertically. In essence, this is a folded column. Blueprint storage rating is totally vertical, while data processing at the opposite extreme is completely horizontal. In the

FIGURE 8-2

Space utilization calculations—elementary school. (From James J. O'Brien (ed.), *Scheduling Handbook*, McGraw-Hill, New York, 1969.)

$$\text{School utilization} = \frac{\text{room utilization} \times \text{unit}}{\text{units}}$$

Config-uration	Units	Comparison, %	Utilization factor			
			Class	Art/Music	Gym	Cafeteria
A	24	100	$\dfrac{77.5(12) + 30(2) + 30(2) + 37\frac{1}{2}(4) + 0(4)}{24} = 50\%$			
B	20	$83\frac{1}{2}$	$\dfrac{77.5(12) + 30(2) + 30(2) + 37\frac{1}{2}(4)}{20} = 60\%$			
C	18	75	$\dfrac{77.5(12) + 30(2) + 67\frac{1}{2}(4)}{18} = 73.5\%$			
D	17	71	$\dfrac{77.5(12) + 60(1) + 67\frac{1}{2}(4)}{17} = 77.8\%$			
E	16	67	$\dfrac{82.5(12) + 67\frac{1}{2}(4)}{16} = 79\%$			

(a)

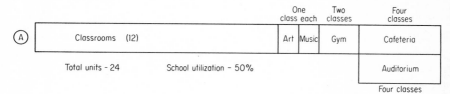

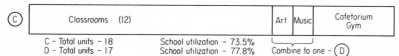

(b)

Classroom utilization.

$$\text{Room utilization} = \frac{\text{weekly use} \times \text{classes}}{\text{classroom size} \times \text{possible sessions}}$$

Configuration	Classrooms	Music/art	Gym	Cafeteria	Auditorium
A	$\dfrac{(40 - 9) \times 12}{1 \times 40 \times 12}$ 77.5%	$\dfrac{2 \times 12}{2 \times 40}$ 30%	$\dfrac{2 \times 12}{2 \times 40}$ 30%	$\dfrac{5 \times 12}{4 \times 40}$ $37\frac{1}{2}\%$	Varies 0
B	77.5%	30%	30%	$37\frac{1}{2}\%$	Combined
C	77.5%	30%	$67\frac{1}{2}\%$	Combined	Combined
D	77.5%	60%	$67\frac{1}{2}\%$	Combined	Combined
E	$\dfrac{40 - 7}{40} =$ 82.5%	None	$67\frac{1}{2}\%$	Combined	Combined

(c)

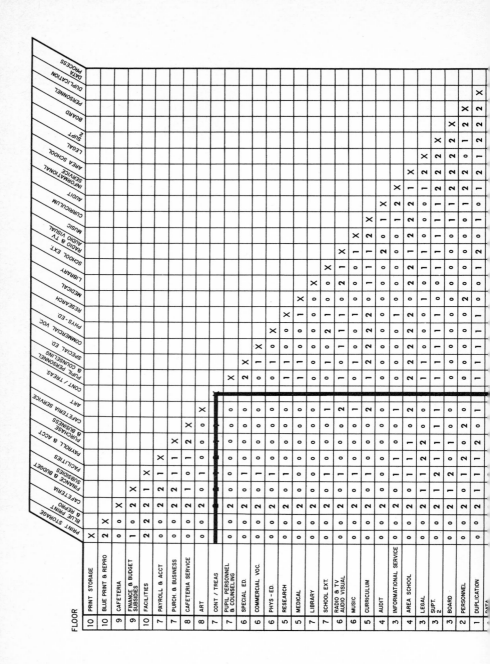

FIGURE 8-3
Adjacency matrix. (From James J. O'Brien (ed.), *Scheduling Handbook,* McGraw-Hill, New York, 1969.)

case of an existing structure, the scoring can be utilized to determine groups which could best move away from the central group. The scores are shown in the totals in the bottom column.

In utilizing a scoring of this type, the unilateral nature should be recognized. Each department has been rated as in close contact with the cafeteria. However, this rating could be much lower if dining facilities were available in the proximity of the building or if the dining facilities were noncentralized.

Figure 8-4 is a stacking chart for the controller-treasurer function. This was developed by totaling the proximity scoring for each floor, extracting the information from the adjacency matrix (Fig. 8-3). Review of the stacking chart indicates that this function should be located either on the seventh to ninth floors, or the first to the third.

In the industrial field, a number of programs have been developed

	Floor	Score	Functions
	10	1	Facilities (1)
	9	4	Finance and budget (2) Cafeteria (2)
	8	1	Cafeteria service (1)
Controller & treasurer	7	3	Payroll & accounting (2) Purchasing & business (1)
	6	0	
	5	0	
	4	0	
	3	6	Legal (2) Board of Education (2) Superintendent (2)
	2	0	
	1	3	Duplication (1) Data processing (2)

FIGURE 8-4
Vertical stacking chart for controller-treasurer function.
(From James J. O'Brien (ed.), *Scheduling Handbook*,
McGraw-Hill, New York, 1969.)

STEEL STRUCTURAL SELECTION MATRIX

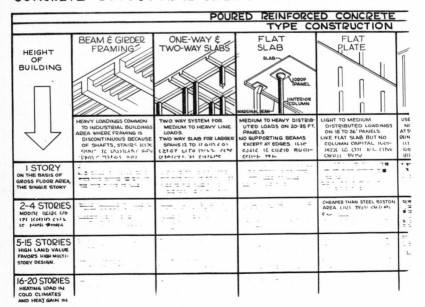

STEEL FLOOR FRAMING SYSTEMS
TYPE CONSTRUCTION

BEAM & GIRDER BEAM & TRUSS	OPEN-WEB STEEL JOISTS	COMPOSITE STEEL-CONCRETE CONSTRUCTION	CELLULAR FLOOR	SHORT SPAN CONCRETE DECK (ON STEEL FRAMING)	STEEL FRAME WITH PRECAST FLOOR PANELS
		A. WITH STUDS (SHOWN ABOVE) B. WITH CHANNELS C. WITH SPIRALS			
1. USE IN CONJUNCTION WITH CONCRETE, STEEL DECK, PRE-CAST PLANK AND ALMOST ALL SLAB SYSTEMS. 2. ADAPTABLE TO FUTURE CHANGES, REINFORCEMENTS AND ALTERATIONS. 3. OPENINGS FOR STAIRS,	1. RAPID ERECTION 2. SUITED TO FRAMED OR WALL BEARING STRUCTURES. 3. CONNECTIONS TO SUPPORT MEMBERS MAY BE WELDED OR FIELD-BOLTED. 4. OPEN WEBS ALLOW PASSAGE OF PIPES, DUCTS & CONDUITS	1. PARTICULARLY USEFUL WHERE AVAILABLE CONSTRUCTION DEPTHS ARE LIMITED, OR FOR INCREASED STIFFNESS 2. COMPOSITE ACTION MUST BE ENSURED BY USING SPECIAL SHEAR CONNEC	1. AVAILABLE IN MANY SHAPES, DEPTHS, THICKNESSES, WITH OR WITHOUT TOP AND/OR BOTTOM PLATES. 2. LIGHT WEIGHT FRAMING SYSTEM. 3. CELLS FORM CONDUITS FOR ELECTRICAL, TELEPHONE	1. FORMWORK FOR SLABS HUNG FROM STEEL BEAMS TO ELIMINATE SHORES. 2. RAPID PLACEMENT OF REINFORCINGS WITH WELDED WIRE FABRIC. 3. ALTERATIONS EASILY	LARGER SPANS & HEAVY LOADS FAVOR PRECAST, PRESTRESSED CONCRETE STRUCTURAL MEMBERS. PRECAST CONCRETE ROOF SECTIONS HAVE THE ADVANTAGE OF RAPID ERECTION, PROVIDE A

CONCRETE STRUCTURAL SELECTION MATRIX

HEIGHT OF BUILDING	POURED REINFORCED CONCRETE TYPE CONSTRUCTION				
	BEAM & GIRDER FRAMING	ONE-WAY & TWO-WAY SLABS	FLAT SLAB	FLAT PLATE	
	HEAVY LOADINGS COMMON TO INDUSTRIAL BUILDINGS AREA WHERE FRAMING IS DISCONTINUOUS BECAUSE OF SHAFTS, STAIRS	TWO WAY SYSTEM FOR MEDIUM TO HEAVY LINE LOADS. TWO WAY SLAB FOR LARGER SPANS 12 TO 11	MEDIUM TO HEAVY DISTRIB-UTED LOADS ON 20-35 FT. PANELS NO SUPPORTING BEAMS EXCEPT AT EDGES	LIGHT TO MEDIUM DISTRIBUTED LOADINGS ON 18 TO 26' PANELS. LIKE FLAT SLAB BUT NO COLUMN CAPITAL	USE
1 STORY ON THE BASIS OF GROSS FLOOR AREA, THE SINGLE STORY					
2-4 STORIES				CHEAPER THAN STEEL BOSTON AREA	
5-15 STORIES HIGH LAND VALUE FAVORS HIGH MULTI-STORY DESIGN.					
16-20 STORIES HEATING LOAD IN COLD CLIMATES AND HEAT GAIN IN					

FIGURE 8-5
Steel and concrete selection matrixes.

to identify functional programs for respective layout of plants. One such program is CORELAP (computerized relationship layout planning). This approach starts with a relationship matrix and generates an experimental physical block plant layout by department.

VALUE ANALYSIS

Throughout the development of the functional program, in particular, there is an opportunity for value analysis and judgments. When the functional program has been completed, a standard value analysis of the high-value components can be conducted. This is actually the first point in the evolution of the project at which this can be done. During budgeting, value analysis is conducted at a higher level, with answers resulting in essentially a go/no-go type of decision. By the time the functional program is completed and costed out, the components have been identified and a cost value assigned for the first time in the project.

In organizations with a value-analysis staff, this value review can be accomplished with an internal group. There is a dual advantage: the availability of the knowledgeable value team and a prerecognition of the utility of value analysis.

The owner may call on outside assistance, even with an internal

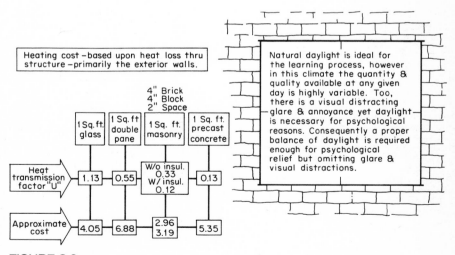

FIGURE 8-6
Transmission and lighting factors. (From an unpublished V.E. study for the Philadelphia School Board; sketch by H. Bryan Loving, R.A.)

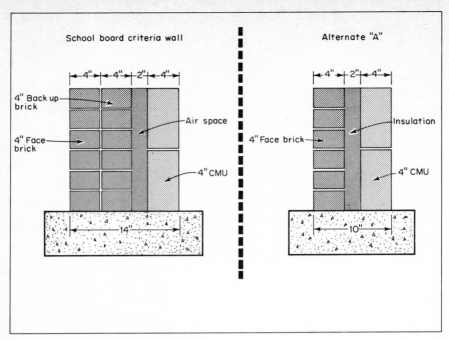

FIGURE 8-7
Wall Study. (From an unpublished V.E. study for the Philadelphia School Board.)

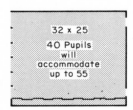

STANDARD RECTANGLE

25 x 32

40 Pupils
will
accommodate
up to 48

1. Provides conventional seating arrangement.
2. Generally provides more natural daylighting, but more visual distraction for pupil, more heat loss due to greater glare.
3. Provides less wall surface for chalkboard of principal teaching location.
4. Results in longer corridors which adds more non-usable area at increase costs.
5. Pupils in last row considerable distance from front teaching area.
6. Greater structural spans and more complicated framing adding to cost.

ROTATED RECTANGLE

32 x 25

40 Pupils
will
accommodate
up to 55

1. Provides less natural daylighting but less pupil distraction, less heat loss due to less glare.
2. Provides more wall surface for chalkboard of principal teaching area.
3. Results in shorter corridors hence less non-usable area = less cost.
4. More pupils closer to teacher.
5. Allows for more informal and non-conventional teaching techniques.
6. Can accommodate more pupils when crowded.
7. Shorter structural span, direct wall bearing lessens costs.

FIGURE 8-8
Room shapes. (From an unpublished V.E. study for the Philadelphia School Board; sketch by H. Bryan Loving, R.A.)

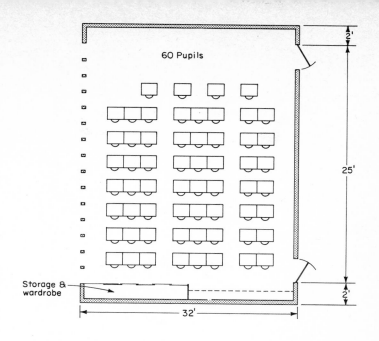

60 Pupils

Storage & wardrobe

25'

2'

2'

32'

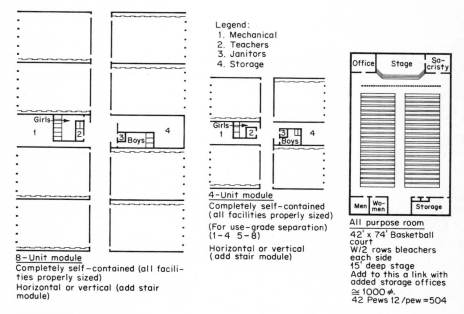

Legend:
1. Mechanical
2. Teachers
3. Janitors
4. Storage

Girls 1 2

3 Boys 4

8-Unit module
Completely self-contained (all facilities properly sized)
Horizontal or vertical (add stair
module)

4-Unit module
Completely self-contained
(all facilities properly sized)
(For use-grade separation)
(1-4 5-8)

Horizontal or vertical
(add stair module)

Office   Stage   Sa-
cristy

Men Wo-   Storage
     men

All purpose room
42' x 74' Basketball
court
W/2 rows bleachers
each side
15' deep stage
Add to this a link with
added storage offices
≅ 1000 ₵.
42 Pews 12 /pew = 504

FIGURE 8-9
Basic planning modules. (From an unpublished V.E. study for the Philadelphia
School Board; sketch by H. Bryan Loving, R.A.)

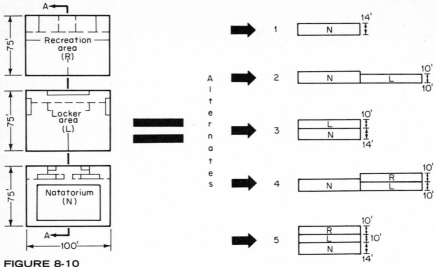

FIGURE 8-10
Alternate arrangements of pool complex. (From an unpublished V.E. study for the Philadelphia School Board.)

team. The GSA does this in contracts where it has assigned a construction manager. Part of the construction manager's role is the conduct of a seminar on value analysis to the design and management team, and application of value techniques to the developing project. However, at the conclusion of the budget and during programming, the design team probably will not have been selected and possibly not even the construction manager.

The programming phase is one which is often omitted. The omission is replaced by bridging the functions of budgeting and design development, resulting in an informal programming. Just as programming is often omitted, value analysis can easily be omitted, even unintentionally in the programming phase. The informed owner should take positive steps to ensure appropriate value analysis during this very important stage in the project's evolution.

Figures 8-5 through 8-10 present a selection of the kinds of results that should be expected of programming phase studies.

CASE STUDY F
PROGRAMMING (HOSPITAL
SURVEY COMMITTEE)

The Hospital Survey Committee of Philadelphia was organized to conduct a survey of hospital facilities in the Philadelphia–South Jersey metropolitan area. It continued on as a viable organization with support from the local business and medical community to evaluate proposed additions to the medical facilities in that area. In this capacity, the Hospital Survey Committee recognized the dramatic acceleration of construction costs in the 1960s. For instance, a new 100-bed hospital which could have been built for approximately $2.5 million in 1960 was priced at almost $8 million in 1969.

In reviewing each individual hospital application, the Hospital Survey Committee was becoming part of a value-analysis review team. It became clearer to the Hospital Survey Committee that better parameters and guidelines for the programming and planning of hospital facilities in the Philadelphia–South Jersey metropolitan area were not available and were necessary for a meaningful evaluation of applications.

The committee commissioned a team made up of hospital consultant John G. Steinle & Associates and the architectural firm of Ewing, Cole, Erdman & Eubank to undertake an integrated study to determine techniques through which the maximum amount of health-care facility construction could be delivered at the lowest possible cost. The effort required a comprehensive study which was funded through the Hospital Survey Committee, and an interim report, *Containment of Construction Costs through Improved Standards and Methods,* was issued in June of 1971. That report contained standards for use in the value assessment of proposed facilities. The initial report also contained an evaluation of project-management techniques in programming design and construction which would permit more rapid delivery of hospital construction. The final phase of the study investigated methods of financing hospital construction, a major cost factor.

The final report, *Cost Containment and Financing of Hospital Construction,* was issued in 1974. It is organized into the following major areas:

Development of space criteria

Productivity of employees in key services

Programming design and construction

Financing needed capital projects

The study notes that if all the recommendations and findings developed had been utilized over the past 10 years, the cost of construction would have been 30 to 40 percent lower. Stated differently, if a hospital board were to adopt the findings of the team and incorporate them into a project of 100 beds, the project would be open in 4 years at a cost of $13.5 million, while a traditional project would open in 6 years at an estimated cost of $15 to $16 million. The Hospital Survey Committee projects a potential for savings of $70 million over the next decade if all the recommendations are implemented.

GUIDELINES FOR SPACE DETERMINATION

While all phases of this report are significant in the overall implementation of hospital projects, this case study focuses upon the physical program parameters developed for the evaluation of space criteria. The following information is extracted from the summary report issued by the Hospital Survey Committee in 1974.

The results of the statistical analysis shown in Table F-1 reflect a variety of relationships ranging from highly significant to nonexistent. For example, the space allotted to outpatient clinics, not surprisingly, was found to bear no noticeable relationship to the bed size and a significant relationship to the workload as reflected in the number of visits. Certain services such as the laboratory exhibited a significant relationship between size and both the workload and the bed complement of the hospital. Space allotted to other functions such as recovery, administration and central service showed only marginal relationships to one or the other variable. Others such as housekeeping, medical records, and the pharmacy showed no relationship whatever to either of the two variables studied, thus suggesting other factors as predominant in the decisions regarding space allocation for those services.

It is particularly noteworthy that certain services noted for the high cost of their construction exhibited significant relationships between space and the workload experienced. Those singled out for further evaluation on this basis were the emergency service, laboratories, radiology, surgery, food preparation service, and cafeteria. As Table F-1 indicates, four of these services exhibited significant relationships between space allocations and both workload and the size of the institution. However, it was deemed important to establish standards on the basis of the workload in view of the fact that most of these services would be increasingly affected by growing outpatient loads; hence, the significance of the relationship between space allocations and bed size would diminish with time.

The final step in the analysis was the development of modules, based on the application of the workload to space ratios to an actual floor plan supplemented by the professional judgment of qualified architects. The procedure involved the preparation by the architects of single-line drawings which identified the various elements required within each module, such as treatment rooms, office space, equipment storage, general storage, utility areas, reception areas, and so forth. Judgmental factors were applied where it was felt that technology, which had not been a factor at the time the hospitals sampled had been designed, could be expected to materially change the amount of space required for a given workload. In addition, each module

Table F-1
Relationship between Space Allocation by Department and Workload or Hospital Size

Department	Relationship of space to:	
	Workload	Total beds
Patient care		
Emergency	Significant	
Inhalation therapy		Significant
Intensive care		Significant
Labor and delivery		Significant
Laboratory	Significant	Significant
Nursery		Significant
Obstetric beds (number)		Significant
Outpatient clinics	Significant	
Pediatric beds (number)		Significant
Physical medicine		Marginal
Radiology	Significant	Significant
Recovery	Marginal	
Surgery	Significant	Significant
Supportive services		
Administration		Marginal
Business office		Significant
Cafeteria	Significant	
Central service		Marginal
Food preparation	Significant	Significant
General stores		Marginal
Housekeeping	None	None
Medical records	None	None
Pharmacy	None	None
Social service		Marginal

SOURCE: *Cost Containment and Financing of Hospital Construction*, 1974 summary report of the Hospital Survey Committee of Philadelphia.

was designed to provide appropriate space for each function within a department, thus resulting in at least minor modifications in actual space allocated as opposed to the theoretical amount required by the application of the preliminary criteria.

The criteria were broadened by designing different modules for workloads of greatly varied size; since doubling the volume to be handled by a depart-

Table F-2
Recommended Criteria Relating Space to Workload, by Service

Service	Ratios			Measurement
	Lower	Median	Upper	
Laboratories	46.88	51.88	56.88	Procedures per square foot
Diagnostic radiology	4.10	4.35	4.60	Procedures per square foot
Surgery	.56	.69	.82	Procedures per square foot
Food preparation	54.00	61.50	69.00	Meals prepared per square foot
Cafeteria	32.50	35.75	39.00	Meals served per square foot
Emergency	6.25	6.60	6.95	Visits per square foot

SOURCE: *Cost Containment and Financing of Hospital Construction*, 1974 summary report of the Hospital Survey Committee of Philadelphia.

ment does not necessarily require a doubling in the size of a department, this too resulted in space allocations somewhat at variance with the theoretical allocations required by the strict application of the tentative criteria. Modules, or architectural models, for three different levels of service volume were designed for each of the six services under study. Subsequently, the ratio of the planned workload to the resulting space allocations was recalculated and a new range of ratios established as a result. Table F-2 lists the resulting ranges and their midpoints, which serve as the criteria recommended for determining the future space needs for the six services covered. In each case, departmental footage involved is expressed in terms of intradepartmental area, that is, all internal gross square footage, including walls, other bearing surfaces, corridors, plus half of all corridor space shared with other departments. The ratios are based on annual workloads, which is true for other ratios cited in this report relating space allocations to workloads.

Additional Guidelines for Space Determination

In addition to the six services singled out for detailed attention, the consultants found a number of other functional areas within the hospitals in which the space allocations bore positive relationships to one or the other of the two variables considered. These functional areas, the variable to which space appeared to be related, and the mean ratio found among the national hospitals studied are indicated in Table F-3.

The ratios shown in Table F-3 are the unrefined averages obtained from the sampling of national hospitals. No effort has been made in the study to fur-

Table F-3
Suggested Guidelines Relating Space to Workload or Hospital Size

Department	Variable related to space allocation	Mean ratio of national hospitals
Outpatient clinics	Workload	3.42 visits per sq ft
Inhalation therapy	Total beds	2.04 sq ft per bed
Labor and delivery	Total beds	13.66 sq ft per bed
Nursery	Total beds	9.83 sq ft per bed
Recovery	Workload	4.95 patient hours per sq ft
Physical medicine	Total beds	12.36 sq ft per bed
Central service	Total beds	7.95 sq ft per bed
General stores	Total beds	28.48 sq ft per bed

SOURCE: *Cost Containment and Financing of Hospital Construction,* 1974 summary report of the Hospital Survey Committee of Philadelphia.

ther develop these into ideal ranges, but they are deemed sufficiently representative to be applied as guidelines in the development of new capital programs. Accordingly, the Hospital Survey Committee recommends the use of these ratios as general guides in the allocation of space for new, expanded, or renovated facilities.

POTENTIAL SAVINGS IN CONSTRUCTION COSTS

The potential impact of the criteria relating space to workloads in the six vital services can be appreciated through an examination of the savings in construction costs which would have resulted from their use in the construction of existing Philadelphia area hospitals. Table F-4 compares actual space allocations in hospitals studied by the consultants with total space needed through application of the criteria.

It will be noted that these figures are based on as few as 15 and at most 27 of the 73 general short-term hospitals in the Philadelphia–South Jersey metropolitan area. It is estimated that the dollar savings for these six functional areas for all 73 hospitals would have been between $25 million and $30 million.

The consultants estimate that the application of these criteria to the development of new hospitals can save 8 to 10 percent of initial construction cost.

SPACE ALLOCATIONS AND SPECIAL INTERESTS

Generally speaking, the services given undue priority are those costing the most to construct. They also tend to be the services supervised by medical

Table F-4
Summary of Space Allocations and Potential Savings in Construction Costs through Application of Workload to Space Criteria for Key Services, Sample Metropolitan Area General Hospitals

Department	Number of hospitals	Present area (sq ft)	Needed area	Potential space savings	Potential cost savings
Laboratories	27	123,525	66,261	57,264	$3,940,000
Radiology	27	149,554	123,903	25,651	1,820,000
Surgery	24	167,018	136,128	30,890	2,750,000
Food Preparation	15	79,148	63,344	15,804	820,000
Cafeteria	15	40,441	34,533	5,908	280,000
Emergency	25	72,953	57,815	15,138	1,240,000
					$10,850,000

SOURCE: *Cost Containment and Financing of Hospital Construction,* 1974 summary report of the Hospital Survey Committee of Philadelphia.

staff officials. Further investigation and analysis disclosed that the influence and pressure for space in surgery, radiology, and pathology almost always originate with physicians, particularly chiefs of service, who constitute probably the strongest special-interest group making unusual demands for space in certain departments or areas.

Because of the prestige of their positions, chiefs of service (whether salaried or nonsalaried) do tend to become very powerful forces in the planning of building programs. However, a special survey of 80 hospitals undertaken by the staff of the Hospital Survey Committee revealed that the majority of chiefs do not tend to remain in office for extended periods of time. Of 69 hospitals replying, only 29 reported that they had retained the same chief of surgery during the 1960s and only 36 of those replying had kept the same chief of anesthesiology. Chiefs of pathology changed at least once during the period at 28 hospitals, and 30 hospitals had had two or more chiefs of radiology.

Establishment of objective criteria to relate workloads to space allocations for major hospital services can minimize the effect of such influences and pressures.

PRODUCTIVITY OF EMPLOYEES IN KEY SERVICES

Investigators studied the volume of procedures or services provided per employee for hospitals of various sizes. It was hoped that this evaluation would pinpoint differences in productivity attributable to economies of

Table F-5
Productivity per Employee for Selected Services, by Hospital Size

Hospital size	Surgery proc/emp/wk	Radiology proc/emp/wk	Laboratories proc/emp/wk	Emergency visits/emp/wk
0–100	4.7	34.2	136.1	26.9
101–200	5.0	45.6	165.0	34.1
201–300	4.5	29.8	178.7	27.2
301–400	5.3	26.6	152.0	32.2
401–500	3.6	17.9	154.2	30.5
501 and over	4.3	20.6	226.4	19.5
Median	4.6	28.2	159.6	28.9

SOURCE: *Cost Containment and Financing of Hospital Construction,* 1974 summary report of the Hospital Survey Committee of Philadelphia.

scale. As Table F-5 indicates, increases in size are not necessarily accompanied by increases in productivity.

The results point to at least one definite trend: hospital size can reach a point beyond which productivity drops in any specialized service. Typically, the smallest hospitals have lower productivity per employee than the next larger size category, but the most productive hospitals may range anywhere from 100 to 400 beds.

Surgery

Productivity trends upward with bed size until the 400-bed limit. The decline beyond that point may be attributable to the relative complexity of procedures performed at the larger hospitals, but it may also reflect a loss of efficiency in these hospitals.

Radiology

Productivity peaks with 200 beds, declining abruptly for larger hospitals. Partly this may stem from the relative complexity of procedures performed in larger institutions, but the steep reduction in productivity suggests an inherent loss of efficiency with an increase in scope of operations.

Laboratory

Productivity of laboratory procedures increases steadily to a size of 300 beds; between 300 and 500 beds the level of productivity falls abruptly for hospitals of larger size, presumably reflecting the effects of both greater sophistication in the procedures performed and a loss of efficiency with increased scope of operations. Even at the markedly lower productivity level exhibited, larger hospitals remain far more productive than the hospitals of under 100 beds in size.

Hospitals with 500 or more beds show a dramatic increase in productivity for laboratory procedures far in excess of the levels attained by hospitals of any smaller size. This probably results from the use by hospitals of this size of automated testing equipment, which at the time of the study was not prevalent among hospitals of smaller size.

Emergency

Productivity per employee in the emergency service appears quite similar for hospitals of varying sizes up to 500 beds. Hospitals of larger size were distinctly lower in productivity, possibly because of the prevalence of teaching programs and a resulting high ratio of personnel to patients in this service. This same factor may affect productivity in the other services mentioned above as well.

COST ESTIMATING

Cost is the principal dimension in value analysis. Without cost for comparison, the analysis of value must necessarily be subjective—and consequently fall short of the full potential.

Cost estimating is a well-developed art in the construction industry. During design, architects and engineers prepare cost estimates for scope control of a project, and during the construction phase for monitoring of the construction and negotiation of change orders. Contractors perform cost estimating in bidding or negotiating for construction work, as a control for profit during the construction, and in the development of change-order proposals and negotiations. The project managers and construction managers utilize cost estimating at all phases of the project.

When used in value analysis, cost estimating is problematic in that the area to be estimated is not fully defined at the time of greatest potential for value analysis. After the design has been completed, the preparation of cost estimates is a relatively straightforward matter, but at this stage of development many of the value alternatives have been precluded and therefore many opportunities for value improvement are lost.

COST CONTENT

Capital Cost

Whatever cost figures are used for the comparison between value in value analysis, they must be comparable and compatible.

The most common figure used to describe the cost of a project is the construction cost or contractor's cost. Where there is one general con-

Table 9-1

Project	Relative cost, % CC
Budgeting and programming	2
Site acquisition	10
Design	7
Project management	4
Construction cost	100
Changes in scope	5
Interest during construction	12
Furnishings and move-in	10
	150

tract, this would be a single figure. Where there are separate prime contractors, the contractor's cost would be the aggregate of all contract prices including: site work; general construction; structural costs; mechanical, electrical, and plumbing costs; elevator contract; HVAC; and special contracts.

Many important considerations hinge upon the contract cost. One of the most important is the designer's fee, which is often an agreed-upon percentage of the expected contract cost or of the actual cost, whichever is lower. Project costs must also be considered. Table 9-1 represents a range of reasonable project costs relative to construction costs (CC).

Thus, it is clear that the cost of this project is at least 50 percent more than the basic construction costs. In a private venture, additional costs would include taxes during construction, insurance during construction, finder's fee and permanent loan fees, legal and closing costs, title insurance, appraisals, and various leasing commissions.

It is most important to look at the total cost picture in early stages such as budgeting and programming when performing value analysis.

Life-Cycle Cost

In buildings, the initial investment is approximately matched over 20 to 30 years by cost of maintaining and servicing the building, including electrical power and other forms of energy. In buildings such as schools and hospitals, the cost of changes and alterations is substantial, often equaling 50 to 100 percent of the original cost.

Study by the General Services Administration indicates that the salary cost of the personnel utilizing the facility over its life span will be about eighteen times the original cost. Accordingly, any investment in the original cost which improves the productivity of the people utilizing the building will be paid many times over. (However, it becomes problematical in regard to who is the recipient of this production improvement.)

Figure 9-1 shows a breakdown of these costs for a typical building. One of the problems in applying life-cycle costing in basic value analysis is first the question of who is to receive the value and second the availability of accurate figures regarding operating and maintenance costs.

If a developer is going to operate the building after he completes it, value to him would include efficient means of energy utilization. If, on the other hand, he intends to sell the building at completion, he will be much less interested in operating costs. If the developer is the owner and will operate the building for his own account, he is particularly interested in the environmental situation insofar as it directly affects the productivity of the people utilizing the building. In cases where the owner is not designing, building, or developing his own facility, he will be well-advised to carefully specify his requirements so that the final result will incorporate the proper level of value without increasing the cost.

Figure 9-2 shows the results of a study in which an engineering group attempted to identify the optimum point for replacement of a

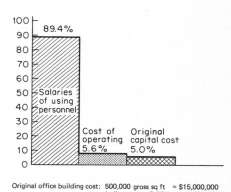

FIGURE 9-1
Breakdown of life-cycle costs for a typical office building.

Original office building cost: 500,000 gross sq ft   = $15,000,000
                                      Indirect costs @ 50% =   7,500,000
                                                                  $22,500,000

Annual operating costs & taxes $2.50/gross sq ft = $1,250,000/year

Annual people cost:

High density/low pay   2,000 people @ $10,000 = $20,000,000
Medium density/medium pay 1,000 people @ $20,000 =   20,000,000

Costs based on 20 year life: Operating $ 25,000,000
                                          Personnel   400,000,000
                                                              22,500,000

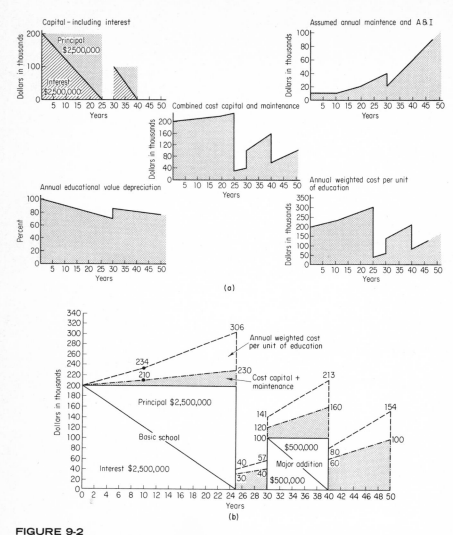

FIGURE 9-2
Summary of elementary school annual facility costs. (a) components of annual school operating cost; (b) annual cost analysis for one elementary school. (From James J. O'Brien (ed.), *Scheduling Handbook*, McGraw-Hill, New York, 1969.)

structure in terms of operating costs. The upper portion shows components of an annual school operating cost. These combined figures are based on the following assumptions:

Total original capital cost, $2,500,000

Bonds, interest 4 percent at 25-year life

Overall period of study, 50 years

$500,000 addition is added at year 30

The purpose of the study was to develop a rational relationship between age and condition of facility to suggest a reasonable range of feasible reinvestment. General rules of thumb utilized indicate that investment of more than 50 percent of replacement costs in terms of additions or renovations to an existing facility should be approached with care. Also, standing procedures indicated that an investment of more than 70 percent of replacement value was not fiscally feasible.

The lower half of Fig. 9-2 shows the combined cost of operating the school. It indicates that the operating costs exceed the original cost only briefly at year 40. If the original capital costs were slightly less per year, the annual maintenance costs were somewhat higher, and the annual depreciation occurred at a more rapid rate, then a breakeven point would be established at which the maintenance of the older facility would exceed the cost of maintaining a new one.

Safety considerations were not introduced. There is such a tendency to continue the use of older buildings that in the Philadelphia school district only the difference in category between fire-safe buildings and non-fire-safe buildings provided a solid guideline for a massive replacement program over the past 10 years.

In order for value analysis to fully utilize the concept of life-cycle costing, better records will have to be maintained on operating costs. Also, means of evaluating depreciation and useful output must be developed.

ESTIMATING METHODS

Estimating the cost of a project, or a portion of a project, is basically a two-step process. The first step is defining the project in terminology to which cost can be applied, then assigning unit cost and extending the unit cost times the number of units in order to arrive at a component cost. The aggregate of these costs is the project cost. Estimating procedures vary considerably from industry to industry, and many special approaches have been developed. The following list is not all-inclusive but does describe the basic well-known approaches.

User Units

The facility to be designed or constructed is defined in terms of its capacity to serve—such as number of patient beds, number of prison inmates, student population, or number of parking spaces. In some cases, a using agency which builds the same type of structure many times has a well-developed historical file on cost and use requirements. An estimate developed on this type of data can be quite accurate.

Square Foot

Square-footage costs are extended by multiplying the quantity of space required times a cost factor developed by either the designer or the owner or more usually by using a square-footage factor developed by organizations such as F. W. Dodge, which has a composite *Dodge Digest* that provides information on more than 2,500 cost analyses for 57 major types of buildings.

Cubic Foot

This approach is very similar to the square-footage procedure except that the cubage provides a measure of the building enclosure size as well as the plan view area.

Parametric Estimating

This procedure involves identifying the scope of the major 15 to 20 subsystems which make up the building and then applying costs to each system based upon historical data or examples of similar buildings.

Modular Costs

In a situation in which the project is made up of repetitious modules, such as housing, garden apartments, and apartments, an accurate cost is developed for one unit and extrapolated by the total number of units and their specific characteristics.

Combined

The combined approach is used when the project is only partially defined. A detailed bill of materials-type takeoff is performed for those portions which are defined, and square-footage costs are used for undefined systems. Most usually, the square-footage cost is applied to the electrical, mechanical, and plumbing sections of the building, which are the last to be defined.

Quantity Survey

This is the European description of a complete takeoff of all materials in the project. Unit prices are applied to each type of material and the results extended and summed up to provide a total cost.

Bidding

This approach involves furnishing a description of a portion of the project to a contractor or supplier specializing in that portion. The job is then costed in the form of a proposal to do the work. This approach can be used either before the project is awarded or in developing costs for extra work.

COST-ESTIMATING STAGES

The method of cost estimating varies with the stage of development of the project. Major project cost decisions must be made in the early budgeting and programming phases. Unfortunately, this is the time when the least is known about the building configuration, dimensions, and scope. Accordingly, the costs which support value analysis at these stages must be similarly limited. Since the cost information available is limited, the estimating techniques utilized are those appropriate to more limited scope information.

As the project becomes better defined, better cost information and estimating techniques are used, and value analysis can be utilized to counter the natural tendency for costs to inflate without increase in value.

In some projects, the cost limitation parameters are absolute. A school district may be mandated by law to spend only a limited amount per square foot of building or for pupils served. The design process must recognize this, or there will be no project. Time is another limitation. A world's fair scheduled to open in 1980 will be useless if it opens in 1982, even if the cost should be within budget. The parameters involved in each project are the result of many factors, and these factors in turn provide specific value goals and objectives.

Figure 9-3 shows the primary approaches utilized at various stages

Project phase	Budget	Program	Schematic	Preliminary	Working documents	Construction
User units	X	X				
Square foot Cubic foot		X	X			
Parametric		X	X			
Module		X	X	X		
Combined			X	X		
Quantity survey				X	X	X
Bids					X	X

FIGURE 9-3
Primary estimating approaches at various stages of project definition.

of project definition. An in-depth survey of estimating approaches utilized in the various design phases of a hospital was conducted for the Veterans Administration by G. Hollander (Ref. 1). The survey determined that dimensional or square-foot methods were used by 82 percent of the firms to prepare budget estimates. It was interesting to note that 53 percent of the firms used more than one approach to check the validity of the original estimate. In preparing the final or preconstruction estimate, almost all respondents indicated reliance on a quantity takeoff. At this stage, only 19 percent used a check estimate.

Budget Estimate

This estimate is prepared in response to a demonstrated need. Accordingly, it should best be related to the number of people to be served, units to be produced, or functions to be accomplished. The best source is prior experience. A multiproject builder such as the Veterans Administration can call upon its own files and experience. A study of 22 VA hospitals built between 1960 and 1973 is described in Ref. 1. The study utilized a statistical regression and correlation technique to adjust the bid price for each project to reflect escalation so that a direct correlation between projects could be prepared.

Gross area, number of beds, and perimeter area of the hospital were used as independent variables. The results were then made available for budgeting future VA hospitals. The one-time builder or inexperienced owner must acquire this type of budgetary information from available reports of comparable experiences of others.

Programming Estimate

At the program stage, the budget has been approved, or at least defined, and it is now converted into some type of facility. User units can still be utilized—but now as a measure of the conversion—dimensional square and cubic prices can be used, and where appropriate, parametric and modular estimates can be introduced.

In addition to basic size factors, factors must be considered for escalation during design and construction, special area costs, and any special conditions or problems. Equipment, furnishings, site costs, and other factors must be included to provide a total estimate. Value analysis is based upon a subbreakdown of the cost factors. In this regard, the parametric approach can be particularly supportive of value analysis.

Schematic Estimate

At the schematic stage, the spatial design has been committed to a plan and a few elevations. There should be a site plan available. To a large degree, this phase utilizes square-footage costs, but parametric and modular costs can become much more practical. It is also possible to segregate certain areas and provide detailed cost-breakdown analyses for value analysis.

A comparative approach can be used for value analysis wherein a segment of work can be assigned an undefined, but common, cost. For instance, if it has been decided that a cafeteria will be located in the building, but the actual location is unknown, the analysis of the cafeteria itself can be deferred; comparative costs can then be assigned to its various locations, leaving the detailed value analysis of the cafeteria to a later phase.

Preliminary Estimate

The preliminary estimate is based upon drawings which are an evolution of the schematic drawings. They are often a full size larger ($^1/_8$ in. = 1 ft, rather than $^1/_{16}$ in. = 1 ft), and therefore accurate quantity takeoff is possible. The spatial solution for the building has been completed, and so the architectural and structural takeoff is substantially accurate. At this stage, the mechanical, electrical, heating, and ventilating systems are just coming into focus. The estimating approach is usually combined, since a quantity takeoff is possible only for architectural and structural, and parametric or square-foot costs must be used for the mechanical and electrical systems.

Value analysis of the preliminary stage has solid cost for the architectural and structural elements. Any low-value areas can be detected. Detailed value analysis for functional areas for services can be conducted. Timeliness is important, since the spatial solution has been defined and major changes become expensive to implement.

Working-Document Estimate

At this stage of project development, a tremendous amount of detail work is in progress. Estimates can be made on a quantity takeoff basis for all areas. The designer will be doing a quantity survey-type estimate, but once the drawings and specifications have been released to contractors, one of the principal methods of developing a cost bid is to solicit firm price quotations from subcontractors. Value analysis now has much more information to work with, but if the project is not to suffer major delays, the results of the analysis must be tailored to details, usually in changes and materials but not configuration. These can still be very significant and important and can be released even during the bidding stage by addendum.

Construction Estimates

Estimates during construction are based on reality. The actual cost of work in place can be used as a guideline in estimating extra work or changes in scope. Quantity survey methods are used by contractors in estimating the projected savings in scope which would result from accepted value-incentive programs.

COST-ESTIMATING SOURCES

Construction companies generally maintain their own file of cost data for use in future estimates. Architects, engineers, and owners not having access to these proprietary files utilize comprehensive sources such as:

1. *Building Cost File* is published on an annual basis by Construction Publishing Co., Inc., New York, and is offered in editions tailored to the specific region of the buyer.

2. *Building Construction Cost Data* is published on an annual basis by Robert Snow Means Co., Inc., Duxbury, Mass.

3. *Dodge Manual for Building Construction Pricing and Scheduling,* is published annually by the Dodge Building Cost Services of the McGraw-Hill Information Systems Company, New York.

4. *Richardson Estimating & Engineering Standards* is published by Richardson Engineering Services, Downey, Calif., on an annual basis.

The regularly published cost indexes have evolved over a number of years and have generally improved with this evolution. The basic information found is contained in a detailed component breakdown commensurate with the manner in which a project is constructed.

Figure 9-4 is one page of the 1975 *Building Cost File* published by Construction Publishing Co., Inc., New York. This book is arranged by Construction Specification Institute (CSI) index numbers for divisions 1–16. Within these categories, concrete is division 3 and heavyweight aggregate concrete is 0331. The costs shown are per cubic yard of concrete in place. They do not include the preparatory work of forms and reinforcing steel. In Fig. 9-5 is an explanation of the method by which a cost estimate is built up using these figures. Note that the qualification is made that contractor's overhead and profit is not a standard figure and must be estimated for each contractor, project, and location. The figures shown are for the New York area. The book includes a conversion table for the conversion of labor and material costs for cities across the country.

Figure 9-6 is a similar information page from the 1974 *Dodge Manual for Building Construction Pricing and Scheduling* by the McGraw-Hill Information Systems Company. This page describes the work involved in assembling form work for concrete. This manual provides derivation of costs, as shown in Fig. 9-7. Even with the changing labor or materials rates, the cost information can be updated for time, escalation, or location by inserting the proper units in the derivation section. For convenience, however, up-to-date manuals should be on hand.

The figures shown are basic labor, material, and equipment costs. They have not been adjusted to include overhead, profit, or special costs.

Figure 9-8 is a page from the *Building Construction Cost Data 1974,* published by Robert Snow Means Co., Inc., of Duxbury, Massachusetts. This page shows the cost of cast-in-place concrete for special categories.

Figure 9-9 shows a typical general contractor's overhead buildup, taken from *Building Construction Cost Data.*

In Fig. 9-8, a Crew C-10 is referred to. Figure 9-10 shows a portion of the typical or standard crew makeup categories, including Crew C-10.

UCI	DESCRIPTION	UNIT	TOTAL	LABOR	MATERIAL	EQUIPMENT
03300	CAST-IN-PLACE CONCRETE					
03311	CONCRETE 2500 PSI					
01	WALL FOOTINGS	CY	48.82	10.13	38.69	
02	COLUMN FOOTINGS	CY	53.97	15.19	38.78	
03	PILE CAPS	CY	55.94	16.93	39.01	
04	TIE BEAMS	CY	54.86	16.02	38.84	
05	GRADE BEAMS	CY	54.86	16.02	38.84	
06	WALLS BELOW GRADE	CY	55.57	16.81	38.76	
07	COLUMN PIERS & PILASTERS	CY	56.28	17.62	38.66	
08	SLABS ON GRADE UP TO 6 INCH THICK	CY	50.79	12.16	38.63	
09	SLABS ON GRADE 7 TO 12 INCH THICK	CY	48.70	10.10	38.60	
10	SLABS ON GRADE 13 TO 24 INCH THICK	CY	46.78	8.20	38.58	
11	SLABS ON GRADE 25 TO 36 INCH THICK	CY	45.93	7.05	38.88	
12	STEPS ON GRADE	CY	55.28	16.72	38.56	
13	PIT SLABS	CY	52.81	13.76	39.05	
14	PIT WALLS	CY	55.75	16.86	38.89	
15	ENTRANCE PLATFORMS ON GRADE	CY	51.11	12.22	38.89	
16	PADS & EQUIPMENT BASES	CY	51.64	12.90	38.74	
17	LOCKER BASES	CY	51.83	12.95	38.88	
18	SLAB EDGED CURBS	CY	55.03	16.07	38.96	
19	COLUMNS	CY	60.21	21.35	38.86	
20	WALLS ABOVE GRADE & SHEAR WALLS	CY	56.68	17.74	38.94	
21	BUTTRESSES	CY	57.19	18.49	38.70	
22	BEAMS	CY	57.56	18.61	38.95	
23	SUSPENDED ARCHS, FLAT SLABS	CY	54.07	15.19	38.88	
24	SUSPENDED ARCHS, BEAMS & SLABS	CY	54.86	16.02	38.84	
25	SUSPENDED ARCHS, PAN & DOME SLABS & BEAMS	CY	55.66	16.83	38.83	
26	SUSPENDED ARCHS, TUBE SLABS	CY	57.03	17.84	39.19	
27	SUSPENDED STAIRS	CY	57.28	18.52	38.76	
28	STEEL BEAM ENCASEMENTS	CY	57.56	18.61	38.95	
29	STEEL COLUMN ENCASEMENTS	CY	60.21	21.35	38.86	
30	FILL FOR METAL PAN TREADS & PLATFORMS	CY	61.40	22.41	38.99	
31	FILL FOR METAL DECKS	CY	56.06	17.73	38.33	
32	MUD SLABS ON GRADE 2 TO 4 INCH THICK	CY	48.24	11.54	36.70	
33	FLOATING SLABS ON SPECIAL FLOATING DECKS	CY	64.89	18.23	46.66	
34	MAT SLABS ON GRADE 18 TO 60 INCH THICK	CY	43.64	6.70	36.94	
35	FOR 3000 PSI STRENGTH ADD/CY TO MATL COSTS	CY	0.68	0.00	0.68	
36	FOR 3500 PSI STRENGTH ADD/CY TO MATL COSTS	CY	0.78	0.00	0.78	
37	FOR 3750 PSI STRENGTH ADD/CY TO MATL COSTS	CY	0.83	0.00	0.83	
38	FOR 4000 PSI STRENGTH ADD/CY TO MATL COSTS	CY	0.88	0.00	0.88	
39	FOR 4500 PSI STRENGTH ADD/CY TO MATL COSTS	CY	0.98	0.00	0.98	
40	FOR 5000 PSI STRENGTH ADD/CY TO MATL COSTS	CY	1.04		1.04	
41	FOR BUILDING HEIGHTS OVER 100 FEET ADD/CY TO TOTAL COSTS	CY	1.99		1.99	
42	FOR BUILDING HEIGHTS OVER 200 FEET ADD/CY TO TOTAL COSTS	CY	2.60		2.60	
43	FOR PUMPED CONCRETE DEDUCT 10 PCT FROM LABOR COSTS	PCT	3.13		3.13	
03312	CONCRETE ADMIXTURES ADDITIONAL COST TO CONCRETE COSTS					
01	1 PCT CALCIUM CHLORIDE ADMIXTURE	CY	.66		.66	
02	2 PCT CALCIUM CHLORIDE ADMIXTURE	CY	1.27		1.27	
03	HIGH EARLY CEMENT ADMIXTURE	CY	1.31		1.31	
04	ANTI-HYDROLITHIC ADMIXTURE	CY	3.91		3.91	
05	WINTER CONCRETE ADMIXTURE	CY	.86		.86	
06	WHITE CEMENT	CY	34.74		34.74	
07	RETARDER	CY	.96		.96	
08	AIR ENTRAINED ADMIXTURE	CY	.52		.52	
03313	CONCRETE FLOOR & STAIR FINISHES ADDITIONAL COST TO CONCRETE COSTS					
01	WOOD FLOAT	SF	.20	.20		
02	STEEL TROWEL, HAND, ONE PASS	SF	.26	.26		
03	STEEL TROWEL, HAND, TWO PASSES	SF	.37	.37		
04	BROOM	SF	.22	.22		
05	MACHINE TROWELED, NORMAL	SF	.21	.13		.08
06	MACHINE TROWELED, VERY DENSE	SF	.17	.12		.05
07	HAND SCREED	SF	.11	.11		
08	VIBRATORY SCREED	SF	.06	.05		.01
09	HARDENER, METALLIC	SF	.38	.19	.19	

1975 BUILDING COST FILE - EASTERN EDITION

FIGURE 9-4

Sample page of *Building Cost File.* (From the 1975 *Building Cost File,* Eastern Edition, Construction Publishing Co., Inc, New York. Used with permission.)

SAMPLE ESTIMATE

A sample estimate has been included below to illustrate how the 1975 BUILDING COST FILE can be used. The particular example chosen shows how prices can be taken from the book and combined into a price per square foot for floor slabs. The source of each price in the book and the method used to convert other units to square feet are noted. A square foot cost analysis is, of course, not typical of all estimating formats. This specific example was chosen as one of the more complex analyses made possible by the data in the FILE.

The general contractor's overhead and profit percentage used is *not* a recommended figure. This percentage will vary in every city, with every contractor, and on every job.

Question: Price a typical designed reinforced concrete flat slab

	UNIT	TOT	LAB	MAT	EQUIP
1. Concrete (10" thick) (Page 41 UCI #0331123) Susp. Arches-Flat slabs	C.Y.	$ 54.07	$ 15.19	$ 38.88	
Convert to $/SF as follows: Price/C.Y. x.83 =	S.F.	1.66	.46	1.20	
27					
2. Formwork (Page 35 UCI #0312302 Susp. Arch Forms (Flat Slab) -3 uses	S.F.	5.59	4.92	.67	
3. Reinforcing (Page 39 UCI #0321609 In Footings & Slabs;	T	970.16	302.27	625.08	42.81
Convert to $/SF: Assume 5 lbs./SF, therefore Price/Ton x 5 =	S.F.	2.43	.76	1.57	.10
2,000					
4. Finishing (Page 41. UCI #0331303 Conc. Finishes; Steel Trowel, Two Passes-Hand)	S.F.	.37	.37		
5. Finishing (Page 41 UCI #0331309 Conc. Finishes; Hardener, Metallic)	S.F.	.38	.19	.19	
Subtotal composite for 10" Reinf. Conc. Flat Slab (Subcontractor's Price—overhead and profit included)	S.F.	10.43	6.70	3.63	.10
For illustrative purposes add 7% for General Contractor's Mark-Up	S.F.	.73	.47	.26	—
Grand Total—General Contractor's price or quote	S.F.	$ 11.16	$ 7.17	$ 3.89	$ 0.10

FIGURE 9-5

Explanation of the method of building up a cost estimate. (From the 1975 *Building Cost File,* Eastern Edition, Construction Publishing Co. Inc., New York. Used with permission.)

CONCRETE FORMWORK   3

L I N E	DESCRIPTION	OUTPUT			UNIT COSTS		
		CREW	PER DAY	UNIT	MATERIAL	LABOR	TOTAL
	** FORM ACCESSORIES (CONT'D) **						
	WALL FORM TIES, (CONT'D)						
1	5000# CAPACITY 8" WALL			EACH	0.21		0.21
2	12" WALL			EACH	0.25		0.25
3	16" WALL			EACH	0.27		0.27
4	24" WALL			EACH	0.31		0.31
5	CONES FOR WALL TIES 1" PLASTIC			EACH	0.08		0.08
6	1 1/2" PLASTIC			EACH	0.10		0.10
7	2" WOOD			EACH	0.15		0.15
	FOR STAINLESS STEEL TIES MULTIPLY COST BY 3.						
8	WALER WEDGES 3000# CAPACITY			EACH	0.31		0.31
9	5000# CAPACITY			EACH	0.55		0.55
	RENTAL (WEDGES) 10% 1ST MO. 5% EACH ADDITION						
10	WALL FORM WALER BRACKETS			EACH	1.53		1.53
	RENTAL 10% 1ST MO. 5% EACH ADDITIONAL						
	** FORMWORK IN PLACE **						
	FOUNDATION FORMWORK CONSTRUCTED OF PLYWOOD						
	UNLESS NOTED. COSTS INCLUDE STRIPPING						
	COLUMN PIER AND PILASTER FORMS						
11	1 USE	3 CP.1 LA	190	SQ FT	0.68	1.82	2.50
12	3 USES	3 CP.1 LA	225	SQ FT	0.30	1.53	1.83
13	5 USES	3 CP.1 LA	240	SQ FT	0.21	1.44	1.65
	GRADE BEAM FORMS						
14	1 USE	3 CP.1 LA	220	SQ FT	0.50	1.57	2.07
15	3 USES	3 CP.1 LA	255	SQ FT	0.22	1.35	1.57
16	5 USES	3 CP.1 LA	275	SQ FT	0.16	1.26	1.42
	PILE CAP FORMS						
17	1 USE	3 CP.1 LA	180	SQ FT	0.56	1.92	2.48
18	3 USES	3 CP.1 LA	215	SQ FT	0.27	1.61	1.88
19	5 USES	3 CP.1 LA	230	SQ FT	0.18	1.50	1.68
	SPREAD FOOTING FORMS						
20	1 USE	3 CP.1 LA	210	SQ FT	0.55	1.64	2.19
21	3 USES	3 CP.1 LA	248	SQ FT	0.26	1.39	1.65
22	5 USES	3 CP.1 LA	270	SQ FT	0.17	1.28	1.45
23	SLAB ON GRADE EDGEFORMS 6" HIGH	3 CP.1 LA	300	SQ FT	0.25	1.15	1.40
24	6" TO 12"	3 CP.1 LA	320	SQ FT	0.27	1.08	1.35
25	12" TO 24"	3 CP.1 LA	280	SQ FT	0.37	1.23	1.60
26	OVER 24"	3 CP.1 LA	250	SQ FT	0.43	1.38	1.81
	WALL FOOTINGS						
27	1 USE	3 CP.1 LA	230	SQ FT	0.55	1.50	2.05
28	3 USES	3 CP.1 LA	260	SQ FT	0.26	1.33	1.59
29	5 USES	3 CP.1 LA	280	SQ FT	0.17	1.23	1.40
	WALL FORMS BELOW GRADE, TO 8' HIGH						
30	1 USE	4 CP.1 LA	230	SQ FT	0.62	1.90	2.52
31	3 USES	4 CP.1 LA	260	SQ FT	0.30	1.68	1.98
32	5 USES	4 CP.1 LA	270	SQ FT	0.22	1.62	1.84
	8' TO 14' HIGH						
33	1 USE	4 CP.1 LA	210	SQ FT	0.68	2.08	2.76
34	3 USES	4 CP.1 LA	235	SQ FT	0.33	1.86	2.19
35	5 USES	4 CP.1 LA	245	SQ FT	0.26	1.78	2.04
	14' TO 18' HIGH						
36	1 USE	4 CP.1 LA	190	SQ FT	0.75	2.30	3.05
37	3 USES	4 CP.1 LA	215	SQ FT	0.36	2.03	2.39
38	5 USES	4 CP.1 LA	225	SQ FT	0.28	1.94	2.22
	18' TO 22' HIGH						
39	1 USE	4 CP.1 LA	170	SQ FT	0.85	2.57	3.42
40	3 USES	4 CP.1 LA	190	SQ FT	0.42	2.30	2.72
41	5 USES	4 CP.1 LA	200	SQ FT	0.33	2.18	2.51
	WALL FORMS BELOW GRADE, TRUE RADIUS TO 8' HIGH						
42	1 USE	4 CP.1 LA	180	SQ FT	0.75	2.43	3.18
43	3 USES	4 CP.1 LA	230	SQ FT	0.35	1.90	2.25
44	5 USES	4 CP.1 LA	245	SQ FT	0.27	1.78	2.05
	8' TO 14' HIGH						
45	1 USE	4 CP.1 LA	165	SQ FT	0.82	2.65	3.47
46	3 USES	4 CP.1 LA	210	SQ FT	0.38	2.08	2.46
47	5 USES	4 CP.1 LA	225	SQ FT	0.30	1.94	2.24

FIGURE 9-6
Concrete formwork information page. (*1974 Dodge Manual for Building Construction Pricing and Scheduling,* McGraw-Hill Information Systems Company, New York. Used with permission.)

HOW THE PRODUCTIVITY DATA ARE PRESENTED

L I N E	DESCRIPTION	OUTPUT			UNIT COSTS		
		CREW	PER DAY	UNIT	MATERIAL	LABOR	TOTAL

For each Item of Work listed under DESCRIPTION, the DODGE MANUAL presents the size and makeup of the CREW used in calculating the Item cost and the crew's productivity. Where the heading OUTPUT appears, the crew's productivity is presented as a quantity of the UNIT indicated for an 8 hour working day; for instance

CREW	PER DAY	UNIT
2 CP, 1 LA	300	SQ FT

indicates that a crew of 2 carpenters and 1 laborer constructs 300 square feet in an 8 hour day.

For certain of the items found in Divisions 15A, Mechanical-Plumbing, and 15B, Mechanical-HVAC, the heading TEAM DAYS will be found instead of PER DAY for equipment installation. This heading indicates the number of 8 hour working days required by the indicated CREW to perform a single installation of the described equipment; for instance:

L I N E	DESCRIPTION	OUTPUT			UNIT COSTS		
		CREW	TEAM DAYS	UNIT	MATERIAL	LABOR	TOTAL
	** STEAM TERMINAL UNITS **						
	UNIT VENTILATORS, INCLUDING LOCAL PIPING						
1	750 CFM	2 PF	2.3	EACH	748.00	514.46	1.262
2	1000 CFM	2 PF	2.4	EACH	825.00	536.83	1.361
3	1250 CFM	2 PF	2.6	EACH	935.00	581.56	1.516
4	1500 CFM	2 PF	2.7	EACH	1.000	603.93	1.603
5	2000 CFM	2 PF	2.8	EACH	1.150	626.30	1.776

FIGURE 9-7
Productivity breakdown. (*1974 Dodge Manual for Building Construction Pricing and Scheduling,* McGraw-Hill Information Systems Company, New York. Used with permission.)

In Fig. 9-8, there are circled figures at the left-hand margin. Figure 9-11 shows the additional information provided in the Means analysis paragraphs for categories 43, 44, and 45.

COST BY SQUARE FOOT, CUBIC FOOT, AND SYSTEM PARAMETER
In the early stages of a project evolution, there are no definitive plans or specifications against which unit costs can be applied. As previously stated, a basic approach is the assumption of a certain size

CAST IN PLACE CONCRETE cont'd.

	CREW	DAILY OUTPUT	UNIT	MAT.	INST.	TOTAL
CEMENT ADMIXTURES cont'd. Water reducing admixture			Gal.	1 to 3		
(44) CONCRETE, FIELD MIX, FOB forms 2250 psi / 3000 psi			C.Y.	20.20 / 21.35		
(42) CONCRETE, READY MIX 3,000 psi, Boston / U.S. average				24 / 20.35		
(41) 5,000 psi, Boston / U.S. average				27 / 23.20		
For unloading time over 30 minutes, add per 15 minutes / For less than 5 C.Y. deliveries, add for each C.Y. less than 5 C.Y.				5 / 8		
CONCRETE IN PLACE including forms (4 use) and reinforcing steel unless otherwise indicated						
(43) Average for concrete framed building, not including finishing			C.Y.	61	77	138
(46) Average for substructure only, simple design / Average for superstructure only			"	40 / 70	38 / 90	78 / 160
(47) Base, granolithic, 1" x 5" high, straight / Cove	C-10 / C-10	175 / 140	L.F.	.04 / .07	1.39 / 1.73	1.43 / 1.80
Beams, 5 kip per L.F., 10 ft. span / 25 ft. span			C.Y.	78 / 76	134 / 103	212 / 179
(117) Chimney foundations incl. Subs O&P, add to the cost of chimneys p. 171				35	40	75
Columns, square, average reinforcing, 12" square / 16" square				106 / 99	209 / 141	315 / 240
Average reinforcing, 24" square / 36" square				90 / 72	98 / 67	188 / 139
12" square, minimum reinforcing / Maximum				84 / 128	197 / 221	281 / 349
16" square, minimum / Maximum				71 / 128	126 / 156	197 / 284
24" square, minimum / Maximum				59 / 121	82 / 114	141 / 235
36" square, minimum / Maximum				51 / 92	55 / 77	106 / 169
Round, spirally reinf., ave. reinforcing, 16" diam. / 20" diameter				184 / 180	129 / 110	313 / 290
24" / 36"				166 / 119	97 / 59	263 / 178
16" round minimum reinforcing / Maximum				121 / 248	95 / 163	216 / 411
20" round minimum / Maximum				113 / 248	74 / 146	187 / 394
24" round minimum / Maximum				100 / 237	63 / 135	163 / 372
36" round minimum / Maximum				88 / 157	42 / 79	130 / 236
Curbs, 6" x 18", straight / Curb and gutter			L.F.	2.75 / 3	2.75 / 3	5.50 / 6
Elevated slabs, flat slab, 100 psf L.L., 20 ft. span / 30 ft. span			C.Y.	50 / 49	56 / 46	106 / 95
Flat plate, 100 psf L.L., 15 ft. span / 25 ft. span				60 / 49	67 / 44	127 / 93
Waffle construction, 24" domes, 100 psf L.L., 20 ft. span / 30 ft. span				64 / 61	67 / 54	131 / 115
One way joists, 20" pans, 100 psf L.L., 15 ft. span / 25 ft. span				74 / 72	102 / 83	176 / 155
One way beam & slab, 100 psf L.L., 15 ft. span / 25 ft. span				66 / 76	93 / 94	159 / 170
Two way beam & slab, 100 psf L.L., 15 ft. span / 25 ft. span				71 / 64	94 / 70	165 / 134

FIGURE 9-8

Cost data for cost-in-place concrete. (*Building Construction Cost Data 1974*, Robert Snow Means Company, Inc. Duxbury, Mass. Used with permission.)

The table below shows a contractors overhead in two ways. The figures on the right are for the overhead mark-up based on both material and labor. The figures on the left are based on the entire overhead applied only to the labor. This figure would be used if the owner supplied the materials or if the percentage of material cost for a particular trade were very low (Plasterers, for instance).

Items of General Contractors Indirect Costs	% of Direct Costs	
	As a mark-up of labor only	As a mark-up of both material and labor
Field Supervision	6.5%	2.6%
Main Office Expense (see details below)	9.2	7.7
Tools and Minor Equipment	1.3	0.5
Workmens Compensation & Employers Liability See ④	5.1	2.1
Field Office, Sheds, Photos, etc.	2.3	0.9
Performance Bond 0.5 to 1.0% Average See ⑦	0.7	0.7
Unemployment Tax See ⑥ (Combined Federal and State)	4.0	1.6
Social Security and Medicare (5.85% of first $12,600)	5.9	2.3
Sales Tax – add if applicable 48/80 x % (Only 4 states	–	
do not have sales tax but project may be exempt)		
Sub Total	35.0%	18.4%
*Builders Risk Insurance See ⑤	0.3	0.3
*Public Liability ''	0.5	0.5
Grand Total	35.8%	19.2%

*Paid by Owner or Contractor

MAIN OFFICE EXPENSE

A General Contractor's main office expense consists of many items not detailed in the front portion of the book. The percentage of main office expense declines with increased annual volume of the contractor. Typical main office expense ranges from 20% to 2% with the median about 7.2% of total volume. This equals 7.7% of direct costs.

The following are approximate percentages for different items usually included in a General Contractor's main office overhead. Different accounting procedures utilized may vary the results from these indicated. Meansco Form 112, Project Overhead Summary, is often used as a check list for these items.

	Typical Range	Average
Managers, clerical and estimators salaries	40% to 55%	48%
Profit Sharing, pension and bonus plans	2 to 20	12
Insurance	5 to 8	6
Estimating and project management (not incl. salaries)	5 to 9	7
Legal, accounting and data processing	0.5 to 5	3
Automobile and light truck expense	2 to 8	5
Depreciation of overhead capital expenditures	2 to 6	4
Maintenance of office equipment	0.1 to 1.5	1
Office rental	3 to 5	4
Utilities including phone and light	1 to 3	2
Miscellaneous	5 to 15	8
	Total	100%

FIGURE 9-9
Typical general contractor's overhead buildup. (*Building Construction Cost Data 1974*, Robert Snow Means Company, Inc., Duxbury, Mass. Used with permission.)

building and the application of comparative cost figures against that assumption. In Fig. 9-12, a page from the *Dodge Digest of Building Costs and Specifications* illustrates a method of presenting cost information on projects. An approach would be to locate a project or group of projects similar to the one to be estimated and then escalate costs to projected time of construction for comparison. Where the match is a good one, this method can get good results, although the description is so brief that it cannot introduce all factors or unusual situations which the actual price may have reflected.

In Fig. 9-13 is an extract of square-footage costs for typical buildings from the Means *Building Construction Cost Data.* In most cases,

STANDARD CREWS Cont'd.

Crew No.	Base Costs		Incl. Subs O&P	
Crew C-10	Hr.	Daily	Hr.	Daily
1 Building labor	$ 7.15	$ 57.20	$ 9.65	$ 77.20
2 Cement Finishers	9.45	151.20	12.55	200.80
2 Gas eng. power tool		34.40		37.80
24 M.H., Daily Total		$242.80		$315.80

Crew C-11	Hr.	Daily	Hr.	Daily
4 Structural Steel workers	$10.30	$329.60	$15.05	$481.60
1 Cement Finisher	9.45	75.60	12.55	100.40
1 Crane operator (heavy)	10.10	80.80	13.95	111.60
1 Oiler	8.85	70.80	12.20	97.60
1 40 ton truck crane		219.00		240.90
56 M.H., Daily Total		$775.80		$1032.10

Crew C-12	Hr.	Daily	Hr.	Daily
2 Cement Finishers	$ 9.45	$151.20	$12.55	$200.80
1 Crane operator (medium)	9.75	78.00	13.45	107.60
1-25 ton hydraulic crane		212.00		233.20
24 M.H., Daily Total		$441.20		$541.60

Crew C-13	Hr.	Daily	Hr.	Daily
1 Structural Steel worker	$10.30	$ 82.40	$15.05	$120.40
1 Welder	10.20	81.60	14.90	119.20
1 Carpenter	9.60	76.80	12.85	102.80
1 Electric welder		10.00		11.00
24 M.H., Daily Total		$250.80		$353.40

FIGURE 9-10
Standard crew makeup categories. (*Building Construc-
tion Cost Data 1974*, Robert Snow Means Company, Inc.,
Duxbury, Mass. Used with permission.)

there is a separate breakdown for the major subcontracts, which can be particularly useful in preparing a combined estimate during the schematic and even preliminary phases.

A number of organizations began an in-depth analysis of historical cost factors. One of these, *Engineering News-Record* (ENR), described the use of parametric costs as an aid in building estimating (Ref. 2). This article noted that parametric cost analyses had appeared in 18 previous ENR quarterly cost reports. (These reports are available from the business data department of *Engineering News-Record.*)

Each project is described in terms of 15 physical building parameters. Other measures such as story height, typical areas, air conditioning capacity, lobby area, and other special conditions are also listed to complete the building definition. A standard listing of 45 trades is used to break out costs for each building.

(43) CONCRETE FOR ENTIRE BUILDING (p. 56 to 63)

These figures give an average cost per C.Y. for all concrete in a typical multi-storied reinforced concrete building. It is assumed that ready mixed concrete is used and that forms will be used 4 times. Floor finish or rubbing are not included; 5% concrete waste is included. When form and reinforcing quantity ratios are available compute the total C.Y. costs from the unit prices in the front of the book. See (12) for handling equipment.

	Material	Labor	Hoist & Equipment	Total
Concrete	$20.35	$ 7.85	$3.50	$ 31.70
Forms	18.75	51.15	.80	70.70
Reinforcing	22.00	12.65	.55	35.20
Total per C.Y.	$61.10	$71.65	$4.85	$137.60

(44) FIELD MIX CONCRETE (p. 58)

Presently most building jobs are built with ready mix concrete except for isolated locations and for some larger jobs requiring over 10,000 C.Y. where land is readily available for setting up a temporary batch plant.

The most economical mix is a controlled mix using local aggregate proportioned by trial to give the required strength with the least cost of material.

With maximum size aggregate of 1-1/2", the cost per C.Y. is tabulated below.

28 day Strength	3000 psi		2250 psi	
Mix Proportions	1:2¼:3		1:2¾:4	
Cement, trucked in bulk @ $1.35 per cwt	6-1/4 bags	$ 7.95	5 bags	$ 6.35
Sand @ $3.50 per ton	.68 tons	2.40	.7 tons	2.45
Stone @ $3.45 per ton	.9 tons	3.10	1.02 tons	3.50
Set up plant, foundations, piping, etc.		.45		.45
Plant, trucks and equipment rental		2.65		2.65
Labor to mix and transport to forms		4.25		4.25
Supervision		.55		.55
Cost per C.Y. FOB Forms		$21.35		$20.20

See (40) for material prices in the major cities.

(45) PLACING READY MIXED CONCRETE (p. 62)

For ground pours allow for 5% waste when figuring quantities.

Prices in the front of the book assume normal deliveries. If deliveries are made before 8 a.m., after 5 p.m. or on Saturdays add $2 per C.Y. Large volume discounts are not included in prices in front of the book.

For the lower floors without truck access, concrete may be wheeled in rubber tired buggies, conveyer handled, crane handled or pumped. Pumping is economical if there is top steel. Conveyers are efficient only for thick slabs. Concrete pump with an operator and one laborer can be rented for $48 per hour for pump with boom, $38 per hour without a boom. Both have a 5 hour minimum rental charge, plus $21 per hour travel time both ways.

At higher floors the rubber tired buggies may be hoisted by a hoisting tower and then wheeled to location. Placement by a conveyer is limited to three floors and is best for high volume pours. Pumped concrete is best when building has no crane access. Concrete may be pumped directly as high as 36 stories using special pumping techniques. Normal maximum height is about 15 stories.

Best pumping aggregate is screened and graded bank gravel rather than crushed stone.

Pumping downward is more difficult than pumping upwards. Placing by cranes, either mobile, climbing or tower types continues as the most efficient method for high rise concrete buildings.

Cost per C.Y. for Wheeled Concrete, dumped only (not incl. screeding or vibrating)							
Item	10 C.F. Walking Cart				18 C.F. Riding Cart		
	Hourly Cost	Wheeled up to 50 ft.	Ea. added 100 feet	Ea. added 5 minutes	Hourly Cost	Wheeled up to 50 ft.	Ea. added 100 feet	Ea. added 5 minutes
Building labor	$7.15	$1.40	$.16	$.72	$7.15	$.69	$.05	$.72
Foreman	.75	.15	.02	.08	.75	.07	.01	.08
Concrete cart	3.50	.70	.08	.35	4.85	.47	.03	.49
Runways, etc.	–	.25	.09	–	–	.25	.09	–
Total Cost/C.Y.		$2.50	$.35	$1.15		$1.48	$.18	$1.29
Hourly production		5.1 C.Y.				10.4 C.Y.		

FIGURE 9-11

Special analyses. (*Building Construction Cost Data 1974*, Robert Snow Means Company, Inc., Duxbury, Mass. Used with permission.)

DODGE DIGEST of BUILDING COSTS and SPECIFICATIONS

ISSUE DATE 3/73

HOSPITALS

NO.	LOCATION	BID DATE	STORIES	STRUCTURE	EXT.	ROOF	FLOOR	INT. WALL	TOTAL SQUARE	TOTAL CUBE	STRUCT.	PLUMBING	H.&V.	A/C	ELECT.	MISC.	TOTAL	$/SQ.	$/CUBE	SPECIFICATIONS
21	Ward Buildings Vernon, TX	12/71	1	stl. frm. & jst. strucl. syst.	face brk.	rigid bd., 3 ply BU	V.A.T., quarry tile	textured drywall ptd.	66,798	831,635	1,170	151	-----	-222--	148	*13	1,704	25.51	2.05	Double "H" - 3 buildings - 120' x 220' -- 300 beds, 12 each 24 bed wards & 12 quiet rms., fire alarm syst. Gas furnaces 3,168,000 BTU/Hr., 9 roof top units htg. syst. roof top 852,000 BTU/Hr. w/90,000 CFM air cond. syst. *Built-in Equip.
22	Hospital Dansville, NY	9/71	3 & part bas.	strucl. stl. strucl. syst.	birch	comp. mtl. dk.	VAT, quarry	RMC	67,000		2,204		----1,084----		630		4,132	61.68		Rectangular - 163' x 192' -- 85 beds, dietary, surgical & x-ray facilities. Water, air htg. syst.; absorption chiller, 6000 lbs. hr. air cond. syst.
23	Hospital Kansas City, MO	9/69	4 & bas.	rein. con. strucl. syst.	face brk., conc. blk. bk-up	flat slab	vinyl asb. tile	gyp-bd.	68,000	808,000	1,660		----1,085----		445		3,180	46.62	3.95	Irregular square - 162' x 161' -- 107 beds, upper flr. not to be fin., mech. & elec. rough-in only.
24	Hospital Cody, WY	8/72	3 & part. bas.	conc. frame struct. syst.	brk.	conc. jst.	VAT, terr., carpet	plast-cer., vinyl fabric	78,640	1,116,590	1,873	284	414	276	375	*117	3,339	42.45	2.99	"Y" shaped - 122' x 197' outside dim. -- 43 bed acute & 40 bed long-term care, medical & food serv. equip. Steam heat. syst. & cap., chilled water - 184 ton air cond. syst. * Built-in Equip.
25	Hospital & Clinic E. Los Angeles, CA	12/69	3 & bas.	conc. frm. pan jst.	slump stone	stl. frm.	cer. tile resil.	cer. tile drywall	95,000								3,918	41.00		"L" shaped - 250' x 220' -- 110 beds 3/full adjunct servs. Warm air heat exch. syst.; elec., 350 tons air cond. syst.
26	Hospital Tucson, AZ	7/69	5 & bas.	stl. frm. strucl. syst.	brk., mas. agg. pnls.	conc. on stl.	vinyl cer. tile	plast. vinyl cov'g	104,000		1,801	290		--620---	495	*146	3,352	32.25		Cross - 120 bed genl. hosp. addn. to exist. facil. Gas fired stm. boilers, forc. warm air htg. syst.; elec.-driven chillers, 480 tons air cond. syst. *Built-in equip.
27	Hospital Norristown, PA	10/69	2 & part. bas.	rein. conc. stl. strucl. syst.	brk.; conc. blk. bk-up	long span stl. jsts.	resil. sheet vinyl	vinyl cov'g	106,700	1,264,000	2,048	646		--916---	528	*150	4,290	40.21	3.40	Rectangle (ancillary) - 150' x 190', cross (nursing) - 46' x 82' ea. wing -- Full serv. 147 bed acute genl. hosp. incls. care & cardiac care units, surg., radiol, obstet. & lg. O.P.D.facils. 2.-250 H.P. oil/gas boilers htg. syst. 525 tons refrig. ducted air systs. *C.S.S., Kitch.
28	Children's Convalescent Hospital Cincinnati, OH	12/71	4	reinf. conc. strucl. syst.	4" face brk.	conc. slab/w tar & gravel	vinyl asb. tile	paint & some epoxy coatings	120,000	1,350,000	2,243	246		--314---	439		3,242	27.02	2.40	Rectangular - 310' x 90' x 45' -- 85 bed convalescent hospital outpatient facilities; staff kitchen & cafeteria; examination & treatment areas; classrooms; 4 hydraulic elev.; Elec., 3,000,000 BTUH htg. syst.; 2 rooftop refrigeration air cond. syst.
29	Rehabilitation Hospital Grand Forks, ND	5/72	4	brg. walls stl. strucl. syst.	brk. & metal wall panels	BU over stl. & conc. decks	cer. & VAT.	plast.pt. pt.	133,500	1,856,540	2,210	562		291	409		3,473	26.01	1.87	2 Rectangles connected by link - 170' x 266' - 88 beds, depts. for admin., & psycho-social, physical & occupational therapy; vocational adjust., & nursing; recreation facilities, food service equip. Hot water/air, central power plant, heating capacity - 50,000,000 BTU cooling cap. - 330 tons AC.
30	Hospital Burke Co., NC	11/69	4	conc. frm. w/pan jsts.	brk. & thin pnls.; conc.blk. bk-up	conc. w/b.u.	VAT, sheet rubber carpet	plast.pt. vinyl cov. spray glz.	142,755		3,591		----1,550----		732		5,872	41.14		"H" shaped - 237' x 191' -- 161 beds w/all ancillary servs. of genl. hosp. incls. laundry. Central type syst., induc. in pat. rms. & dual duct in other areas - 22,360 #/hr. stm. htg. & 887 tons air cond.

FIGURE 9-12
A method of presenting project cost information. (*Dodge Digest of Building Costs and Specifications,* McGraw-Hill Information Systems Company, New York. Used with permission.)

207

S.F., C.F. and % of TOTAL COSTS	UNIT	LOW	1/4	MEDIAN	3/4	HIGH
AUTOMOTIVE SALES cont'd. Percentage of total; Plumbing Heating, ventilating, air conditioning	%	2.2% 5.6%	4.5% 8.1%	5.6% 10.1%	7.2% 12.8%	15.2% 22.1%
Electrical Total: Mechanical & Electrical	"	4.1% 15.0%	9.0% 23.5%	10.7% 27.9%	13.6% 32.0%	24.9% 49.3%
BANKS Total project costs	S.F. C.F.	14.95 .85	34.15 2.25	43.50 3.05	55.15 4.05	95 8.50
Plumbing Heating, ventilating, air conditioning	S.F.	.40 1.40	1.40 3.25	1.90 4.40	2.75 5.75	9.80 16.35
Electrical Total: Mechanical & Electrical	"	1.05 3.15	3.25 8.40	4.35 10.70	5.60 13.65	12.85 31.25
Percentage of total; Plumbing Heating, ventilating, air conditioning	%	1.1% 4.4%	3.8% 8.2%	4.6% 10.3%	5.9% 13.6%	16.0% 24.0%
Electrical Total: Mechanical & Electrical	"	4.3% 10.6%	8.6% 21.9%	10.4% 26.1%	12.3% 29.7%	28.0% 59.2%
BOWLING ALLEYS Total project costs (see also p. 171)	S.F. C.F.	12.75 .85	18.50 1.05	20.25 1.25	25.80 1.35	27 1.75
CHURCHES Total project costs	S.F. C.F.	12.60 .70	23.65 1.50	28.90 1.80	34.75 2.15	85 5
Plumbing Heating, ventilating, air conditioning	S.F.	.25 1.15	1 2.45	1.35 3.35	1.95 4.45	4.60 13.40
Electrical Total: Mechanical & Electrical	"	.50 1.95	1.75 5.30	2.15 6.55	3.10 8.60	11 21
Percentage of total; Plumbing Heating, ventilating, air conditioning	%	0.7% 5.4%	3.6% 9.5%	4.9% 11.8%	6.4% 14.1%	14.0% 23.3%
Electrical Total: Mechanical & Electrical	"	2.8% 10.6%	6.8% 20.6%	8.3% 24.1%	9.9% 27.9%	21.2% 39.2%
CLUBS Fraternal, Y.M.C.A., etc. Total project costs	S.F. C.F.	14.50 .65	20.65 1.40	25.20 1.75	31.15 2.05	55 4
Plumbing Heating, ventilating, air conditioning	S.F.	.55 1.10	1.20 2.60	1.75 3.55	2.50 4.80	6.70 8.10
Electrical Total: Mechanical & Electrical	"	.80 3.25	1.70 5.35	2.30 7.60	2.85 9.50	6.95 18.65
Percentage of total; Plumbing Heating, ventilating, air conditioning	%	2.3% 6.3%	5.2% 11.0%	6.6% 14.0%	8.7% 18.4%	18.4% 25.2%
Electrical Total: Mechanical & Electrical	"	3.2% 12.8%	7.7% 25.9%	9.2% 29.2%	10.6% 34.7%	26.0% 64.6%
CLUBS Country. Total project costs	S.F. C.F.	12.80 1.10	23.55 1.75	28.15 2.35	33.15 2.75	46.75 3.50
Total: Mechanical & Electrical	S.F. %	3.15 20.1%	6.65 27.0%	8.40 30.5%	10.90 33.9%	13.05 40.7%
COLLEGE Classrooms and Administrative Total project costs	S.F. C.F.	13.75 .75	29.10 2	37.70 2.50	46.20 3	82.65 4.60
Plumbing Heating, ventilating, air conditioning	S.F.	.80 1.10	1.45 4.20	2.10 6.50	2.90 7.90	8 13.30
Electrical Total: Mechanical & Electrical	"	1.40 3.50	2.85 8.80	3.90 11.20	5.05 15.70	11.20 30.75
Percentage of total; Plumbing Heating, ventilating, air conditioning	%	1.9% 4.7%	4.2% 13.4%	5.9% 16.5%	7.4% 19.6%	13.2% 25.6%
Electrical Total: Mechanical & Electrical	"	4.7% 13.3%	8.9% 27.3%	10.1% 30.8%	12.2% 36.7%	27.2% 46.1%
COLLEGE Science, Engineering, Laboratories Total project costs	S.F. C.F.	18.35 .60	32.50 2.30	43.40 2.85	49.50 3.50	93 6.25
Plumbing Heating, ventilating, air conditioning	S.F.	1.15 3	2 5.70	3 7.95	4.15 10.20	8 17.75
Electrical Total: Mechanical & Electrical	"	1.60 6.90	3.05 10.15	3.95 13.30	5.50 17.45	11 27.50
Percentage of total; Plumbing Heating, ventilating, air conditioning	%	2.5% 8.2%	5.4% 13.9%	7.8% 17.4%	9.1% 22.6%	16.1% 32.1%
Electrical Total: Mechanical & Electrical	"	5.0% 16.7%	8.5% 30.5%	10.5% 34.7%	12.6% 40.0%	23.4% 54.4%

SF

FIGURE 9-13

Square-footage costs for typical buildings. (*Building Construction Cost Data 1974*, Robert Snow Means Company, Inc., Duxbury, Mass. Used with permission.)

DISTRIBUTION OF COST

Key to the selection of areas of value analysis is the distribution of cost over the various categories making up the building. Figure 9-14 is an example of one format which can be used to list the cost for a building. The distribution or breakdown can be increased if the level of information available is in more depth. The initial figures entered are estimates. As actual bids are received, these can be entered and coded by circling or other indication to show that they are actual rather than estimate.

A list of average building-systems cost per gross sq ft for a private-office high rise was published in *Architectural Record* of June 1974. Taking these costs and spreading them according to the format in Fig. 9-14 gives the equivalent breakdown shown in Fig. 9-15. According to the Pareto distribution, 20 percent of the items should represent 80 percent of the cost. Of the 17 items listed, the four most expensive are:

Superstructures	$ 5.90
HVAC	5.80
Exterior walls	4.60
Electrical	4.00
	———
Subtotal	$20.30

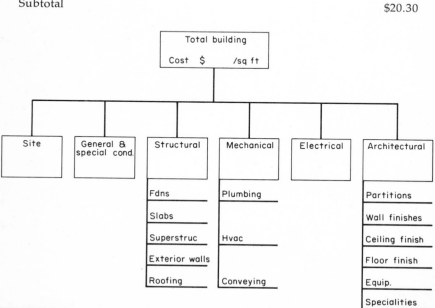

FIGURE 9-14
Cost distribution format.

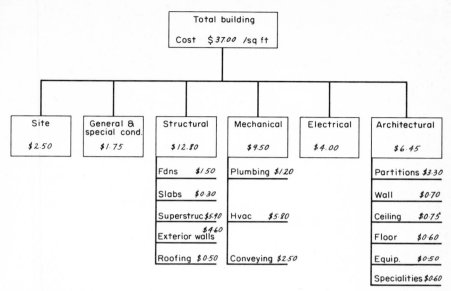

FIGURE 9-15
Breakdown of average building systems cost. (*Architectural Record*, June 1974.)

This indicates that 24 percent of the items equal only 55 percent of the cost. Actually, the rule of thumb is even less effective, since the electrical work can be broken down into power source, distribution system, and fixtures. The HVAC system would break down into heating, ventilating, air conditioning equipment, and distribution systems. The value analyst should still select the high-price categories for review, but the categories should be true functional components, not just a broad disciplinary category. In the model shown in Fig. 9-15, obvious areas for review would be the superstructure at $5.90 and the exterior walls at $4.60.

REFERENCES

1. Hollander, G., *Actual Specifying Engineer,* June 22, 1974, pp. 42–46.

2. "Parameter Costs Aid Building Estimates," *Engineering News-Record,* June 21, 1973, p. 102.

CASE STUDY G
ESTIMATING (AMIS COST MODEL)

The cost estimate shown on the following pages was prepared for a proposed office building in central Florida. (This project was part of a development of a 9-acre site described more fully in Case Study E.)

There were no drawings available, since this estimate was prepared during the feasibility study for the project. The cost system used to generate the estimate in this section was the computerized system of quantity surveying and building-cost evaluation which is a proprietary software system. It was developed by the AMIS Construction and Consulting Services of New York City. This system permits a finite approach to cost-configuration alternates and lends itself very well to value analysis.

The AMIS system is a computer model which accepts basic parameters ranging in square foot per student to dimensions and numbers of stories, percentage occupancy of site, perimeter configurations, glass percentage, and other specific desires of the owner or designer. Given this input, the system generates a design based on the information furnished. Where information has not been furnished, the program has prestored selected features which are reasonable for the type of facility. Page 1 of the output relates this input information or assumption.

The result of the estimate is a raw construction cost which can be factored for area, overhead, location, and type of contract. These factors can be preloaded into the program.

A model can be iterated to generate quantities in cost for various design factors, such as change in the exterior configuration in terms of glass percentage or length-to-width ratios.

The AMIS system provides a means of rapid comparison of alternatives at a time when the owner or designer has the least specific information available. This provides specifics for value analysis, since the model breaks the building down into its components and identifies the areas most suitable for value analysis.

Figures G-1 through G-6 show a computer-generated summary and detailed cost breakdown.

Review of the summary sheet indicates that value and analysis would be best performed in the areas of structural steel, exterior walls, HVAC, or electrical systems. However, review of the detailed breakdown provides more specific identification of the areas with the highest incremental cost. This list would again include structural steel and exterior walls. These items approach $1 million in cost. Items in the half million dollar range would include windows, acoustical tile, plumbing fixtures, fan coil units, and electric light fixtures. The refrigeration equipment is the third most expensive item.

144

BASIC INPUT DATA

SECTOR DATA

(0. VALUE MEANS - DATA NOT AVAILABLE.)

SEC NO.	SECTION NAME	NO OF BLDGS IN SECTION	NET AREA OF SECT.	NO. OF OCCUPNTS	SITE AREA	GROSS/ NET RATIO
1		1	280000.	0.	130680.	1.29

SECTOR NO. 1

BLDG NO.	BLDG NAME	NO OF STORIES	NET AREA	GROSS/ NET RATIO	FLOOR PERIM.	GLASS PCT	BSMNT PCT
1	SINGLE TOWER	25.	280000.	1.29	520.	50.	100.

BLDG NO.	BLDG NAME	COLUMN SPACING	NO. OF COLS.	SLAB LOAD	ROOF LOAD	SOIL/PILE CAPACITY	HVAC SYST.
1	SINGLE TOWER	25.0X20.0	40.	125.	100.	100.	

ASSUMED INPUT DATA

SECTOR DATA

(0. VALUE MEANS - DATA NOT AVAILABLE.)

SEC NO.	SECTION NAME	NO OF BLDGS IN SECTION	NET AREA OF SECT.	NO. OF OCCUPNTS	SITE AREA	GROSS/ NET RATIO
1		1	280000.	0.	130680.	1.29

SECTOR NO. 1

BLDG. NO.	BLDG NAME	NO OF STORIES	NET AREA	GROSS/ NET RATIO	FLOOR PERIM.	GLASS PCT	BSMNT PCT
1	SINGLE TOWER	25.	280000.	1.29	520.	50.	100.

BLDG NO.	BLDG NAME	COLUMN SPACING	NO. OF COLS.	SLAB LOAD	ROOF LOAD	SOIL/PILE CAPACITY	HVAC SYST.
1	SINGLE TOWER	25.0X20.0	40.	125.	100.	100.	FAN COIL

FIGURE G-1

FEDERAL S&L OFFICE BUILDING

SUMMARY

ITEM	UNIT	QUANTITY	UNIT PRICE	TOTAL	GRAND TOTAL
EXCAVATION				30036.	
FOUNDATION CONCRETE				324574.	
STRUCTURAL STEEL				1026502.	
METAL DECK				252839.	
CONCRETE				729631.	
EXTERIOR WALLS				1243554.	
WINDOWS&GLAZING				555100.	
ENTRANCES				45000.	
DAMPPROOF'G & WATERPROOF				34187.	
ROOFING & SHEETMETAL				27558.	
MISC. IRON				251650.	
INTERIOR PARTITIONS				710223.	
SPRAY ON F.P				63629.	
HOLLOW METAL				118450.	
CARPENTRY & MILLWORK				176019.	
HARDWARE				64770.	
FURR&LATH				98885.	
CERAMIC & QUARRY TILES				95572.	
RESILIENT FLOOR				120186.	
ACOUSTIC TILE				372963.	
PAINTING & DECORATING				161060.	
ELEVATORS				584000.	
SITE WORK				196020.	
GENERAL CONDITIONS				436944.	
PLUMBING				595200.	
SPRINKLERS				25300.	
H.V. & A.C.				1900270.	
ELECTRIC				1030320.	
SITE LIGHTING				13068.	
TOTAL COST BID					11283504.

FIGURE G-2

FEDERAL S&L OFFICE BUILDING
PROGRAM BUDGET ESTIMATE

ITEM	UNIT	QUANTITY	UNIT PRICE	TOTAL	GRAND TOTAL
EXCAVATION					
SITE GRADING	S.F.	19492.3	0.10	1949.	
MASS EXCAV. & BACKFILL	C.Y.	6174.3	2.00	12348.	
FOOTING & PIT EXCAVATION	C.Y.	2859.1	3.00	8577.	
BACKFILL	C.Y.	1460.9	3.00	4382.	
POROUS FILL	S.F.	13892.3	0.20	2778.	30036.
FOUNDATION CONCRETE					
CONCRETE IN FOOTGS/CAPS	C.Y.	940.5	80.00	75244.	
CONCRETE IN PIERS	C.Y.	18.6	120.00	2240.	
FOUND. WALLS	S.F.	7354.8	5.25	38612.	
PIT WALLS	S.F.	321.1	5.50	1766.	
PIT SLABS	S.F.	136.0	1.00	136.	
REINFORCING STEEL	TON	64.0	400.00	25613.	
MISC. PADS ETC.	L.S.			14361.	
PILES	L.F.	23800.0	7.00	166600.	324574.
STRUCTURAL STEEL					
STEEL	TON	2211.0	425.00	939675.	
STUDS	S.F.	347307.6	0.25	86826.	1026502.
METAL DECK					
DECK	S.F.	361199.9	0.70	252839.	252839.
CONCRETE					
S.O.G	S.F.	13892.3	1.05	14586.	
FILL ON METAL DECK	S.F.	361199.9	1.05	379259.	
CONC. SPANDRELS	L.F.	13520.0	12.00	162240.	
REBARS	TON	347.0	400.00	138800.	
MISC.	L.S.			34744.	729631.
EXTERIOR WALLS					
EXTERIOR WALLS	S.F.	158600.0	6.00	951600.	
COPING	L.F.	520.0	15.00	7800.	
PERIPHERAL ENCLOSURE	L.F.	13520.0	15.00	202800.	
MISC. WORK	L.S.			81354.	1243554.
WINDOWS&GLAZING					
WINDOWS	S.F.	79300.0	7.00	555100.	555100.

FIGURE G-3

FEDERAL S&L OFFICE BUILDING
PROGRAM BUDGET ESTIMATE

ITEM	UNIT	QUANTITY	UNIT PRICE	TOTAL	GRAND TOTAL
ENTRANCES					
ENTRANCES	S.F.	1500.0	30.00	45000.	45000.
DAMPPROOF'G & WATERPROOF					
DAMPPROOFING	S.F.	7354.8	0.30	2206.	
FABRIC FLASHING	L.F.	13520.0	1.00	13520.	
INSULATION	S.F.	39095.2	0.40	15638.	
MISC. WORK	L.S.			2822.	34187.
ROOFING & SHEETMETAL					
FILL	S.F.	13892.3	0.60	8335.	
BUILTUP ROOF	S.F.	13892.3	1.00	13892.	
FLASHING	L.F.	520.0	5.00	2600.	
MISC. WORK	L.S.			2731.	27558.
MISC. IRON					
STAIRS	EA.	75.0	1250.00	93750.	
LINTELS	L.F.	13520.0	5.00	67600.	
MISC. FRAMING & GRATING	S.F.	361200.0	0.25	90300.	251650.
INTERIOR PARTITIONS					
MASONRY PARTITIONS	S.F.	54179.9	1.90	102941.	
STUDS & DRYWALL	S.F.	307019.8	1.10	337721.	
FURR & DRYWALL	S.F.	187659.9	1.00	187659.	
COLUMNS F.P	S.F.	63000.0	1.30	81900.	710223.
SPRAY ON F.P					
SPRAY ON F.P	S.F.	424199.9	0.15	63629.	63629.
HOLLOW METAL					
EXTERIOR DOORS & BUCKS	EA.	7.0	100.00	700.	
INTERIOR DOORS & BUCKS	EA.	1570.0	75.00	117750.	118450.
CARPENTRY & MILLWORK					
HANG DOORS	EA.	1577.0	20.00	31540.	
PROTECTION & MISC. WORK	S.F.	361200.0	0.40	144479.	176019.
HARDWARE					
EXTERIOR DOORS	EA.	7.0	100.00	700.	
INTERIOR DOORS	EA.	1570.0	40.00	62800.	
MISC. DOORS	L.S.			1270.	64770.

FIGURE G-4

FEDERAL S&L OFFICE BUILDING

PROGRAM BUDGET ESTIMATE

ITEM	UNIT	QUANTITY	UNIT PRICE	TOTAL	GRAND TOTAL
FURR&LATH					
BLACK IRON	S.F.	329619.9	0.30	98885.	98885.
CERAMIC & QUARRY TILES					
C.T. FLOORS	S.F.	17337.5	1.75	30340.	
C.T. WALLS	S.F.	32615.6	2.00	65231.	95572.
RESILIENT FLOOR					
V.A.T FLOORS	S.F.	286276.0	0.30	85882.	
BASE	L.F.	85759.9	0.40	34303.	120186.
ACOUSTIC TILE					
ACOUSTICAL TILE	S.F.	286276.0	1.00	286276.	
METAL PAN	S.F.	43343.9	2.00	86687.	**372963.**
PAINTING & DECORATING					
WALLS	S.F.	801699.6	0.12	96203.	
EPOXY	S.F.	40084.9	0.50	20042.	
DOORS & FRAMES	EA.	1577.0	10.00	15770.	
MISC PAINTING	L.S.			29043.	
ELEVATORS					161010.
ELEVATORS	EA.	8.0	73000.01	584000.	584000.
CLEANING		361200.0	0.00	0.	0.
MISC CONTR.		361200.0	0.00	0.	0.
CABINETS & CASEWORK				0.	0.
SITE WORK					
SITE WORK	L.S.			196020.	196020.

FIGURE G-5

FEDERAL S&L OFFICE BUILDING

PROGRAM BUDGET ESTIMATE

ITEM	UNIT	QUANTITY	UNIT PRICE	TOTAL	GRAND TOTAL
GENERAL CONDITIONS					436944.
SUBTOTAL					7719350.
BOND					0.
OVERHEAD & PROFIT					0.
SUBTOTAL GEN CONSTR.					7719350.
PLUMBING					
FIXTURES & CONNECTIONS	EA.	493.0	1200.00	591600.	
ROOF DRAINS	EA.	6.0	600.00	3600.	595200.
SPRINKLERS					
SPRINKLERS	HDS.	253.0	100.00	25300.	25300.
PLUMBING SITE CONNET.		0.0	0.00	0.	0.
H.V & A.C					
REFRIGERATION EQUIP.	TON	1338.0	600.00	802800.	
FANS	CFM	397399.9	0.30	119219.	
DUCTS	LBS	198699.9	1.50	298049.	
PUMPS	EA.	7.0	2000.00	14000.	
FAN COILS	EA.	676.0	900.00	608400.	
CONTROLS	EA.	289.0	200.00	57800.	1900270.
H.V & A.C UTILITY CONN.		0.0	0.00	0.	0.
ELECTRIC					
MAIN POWER CONN.	KVA.	3612.0	15.00	54180.	
LIGHT&POWER PANELS	EA.	181.0	1000.00	181000.	
LIGHT FIXTURES	EA.	5160.0	80.00	412800.	
SWITCHES	EA.	1290.0	50.00	64500.	
RECEPTACLES	EA.	3612.0	50.00	180600.	
TEL. OUTLETS	EA.	722.0	40.00	28880.	
EQUIPMENT CONNECTIONS	L.S.			108359.	1030320.
SITE LIGHTING					
SITE LIGHTING	L.S.			13068.	13068.
					11283504.
TOTAL					11283504.

FIGURE G-6

The AMIS model is generated without a design prerequisite. It can be maintained and continued during the design phase, with definite specific information introduced into the system as it is available.

The figure generated in this particular estimate is raw and represents the subcontract cost of these items. For an overall single general contract, the total figure of $11,283,504 should be increased to at least $12,186,000.

This cost estimate was generated concurrently with the cost feasibility study for this project. That particular study assumed that this particular office building would have almost 500,000 sq ft and 383,000 net sq ft. Actually, the building estimated had only 280,000 net sq ft, resulting in a lower rentable area. Also, on the basis of 316,200 gross sq ft, the square-foot cost of this building would be $33.74. This figure is in excess of the high range used in the feasibility study. It would be imperative that these figures be reintroduced into the feasibility study to determine whether or not to proceed with the project.

CASE STUDY H
SCHEMATIC DESIGN (COST ESTIMATING)

The following estimate was prepared by the office of James J. O'Brien, P.E., using a schematic design for a science and technology academic building at a New York community college. The estimating approach was a quantity takeoff of the spatial design using a combined approach. Where units were available, costs were assigned and extended. Where sufficient definition was not available, a parametric square-foot approach was used. The information was introduced into a computer system developed by Wood & Tower of Princeton, New Jersey. This system performs a number of functions. Given units of quantity, this system will either input its own unit cost for the type of building and location or will accept an override input by the estimator. The computer-generated cost breakdown shown in Figures H-1 through H-12 includes both approaches. (Where the estimator agreed with the Wood & Tower figure, it remained in, and in situations where the estimator had special knowledge or there were special conditions, an override unit was placed in. In some cases, a lump sum was inserted and so noted.)

Another advantage of the Wood & Tower approach is its ability to sort the information in a number of different configurations. Accordingly, information is available on the system report by major categories and is further broken down into labor and material by major categories. For the value analyst, perhaps one of the most important factors is the square-footage cost by parameter. This permits a searching out and identification of those parameter costs which appear to be high in comparison to usual costs. This then permits the value analyst to focus on the most fruitful areas for analysis. When areas for analysis have been identified, the system-component report is useful in that it presents quantities of material required as well as labor and material. This permits further analysis.

At the preliminary stage, this same program can be used, or an expanded version can be initiated. The expanded version accepts a much broader level of detail. For instance, where the system during schematic would take yards of concrete, at the preliminary or preconstruction estimate the system would take pounds of reinforcing steel, formwork, form accessories, and yards of concrete as a material.

The Wood & Tower system is continuously updated to reflect current labor and material prices so that the estimate generated is current.

SYSTEM SUMMARY

DESCRIPTION	LABOR	MATERIAL	TOTAL	SQ FT
ROADS, WALKS & PARKING	9,587	7,457	17,044	0.10
FOUNDATIONS	181,682	145,256	326,938	1.99
FLOORS ON GRADE	60,205	41,994	102,199	0.62
SUPERSTRUCTURE	544,060	361,722	905,782	5.50
ROOFING	65,355	50,480	115,835	0.70
EXTERIOR WALLS	267,741	1,262,687	1,530,428	9.30
PARTITIONS	236,901	351,128	588,029	3.57
WALL FINISHES	52,265	31,160	83,425	0.51
FLOOR FINISHES	49,135	232,791	281,926	1.71
CEILING FINISHES	29,875	56,175	86,050	0.52
VERTICAL TRANSPORTATION	72,000	108,000	180,000	1.09
SPECIALTIES	17,693	110,103	127,796	0.78
FIXED EQUIPMENT	0	911,245	911,245	5.54
GENERAL CONDITIONS	0	480,000	480,000	2.92
H.V.A.C.	477,340	477,340	954,680	5.80
PLUMBING	325,908	276,528	602,436	3.66
ELECTRICAL	516,844	387,120	903,964	5.49
CONSTRUCTION TOTAL	2,906,591	5,291,186	8,197,777	49.80

FIGURE H-1

SYSTEM REPORT

DESCRIPTION	TOTAL COST
ROADS, WALKS & PARKING	17,044
FOUNDATIONS	326,938
FLOORS ON GRADE	102,199
SUPERSTRUCTURE	905,782
ROOFING	115,835
EXTERIOR WALLS	1,530,428
PARTITIONS	588,029
WALL FINISHES	83,425
FLOOR FINISHES	281,926
CEILING FINISHES	86,050
VERTICAL TRANSPORTATION	180,000
SPECIALTIES	127,796
FIXED EQUIPMENT	911,245
GENERAL CONDITIONS	480,000
H.V.A.C.	954,680
PLUMBING	602,436
ELECTRICAL	903,964
CONSTRUCTION TOTAL	8,197,777

FIGURE H-2

```
FEBRUARY 20, 1974                                              PAGE   1
                      GROUP-SYSTEM REPORT
-----------------------------------------------------------------------
DESCRIPTION                LABOR      MATERIAL      TOTAL    SQ FT
-----------------------------------------------------------------------

SITE IMPROVEMENTS

ROADS, WALKS & PARKING      9,587        7,457      17,044    0.10
                            -----        -----      ------    -----
                            9,587        7,457      17,044    0.10
ARCHITECTURAL & STRUCTURAL

FOUNDATIONS               181,682      145,256     326,938    1.99
FLOORS ON GRADE            60,205       41,994     102,199    0.62
SUPERSTRUCTURE            544,060      361,722     905,782    5.50
ROOFING                    65,355       50,480     115,835    0.70
EXTERIOR WALLS            267,741    1,262,687   1,530,428    9.30
PARTITIONS               236,901      351,128     588,029    3.57
WALL FINISHES             52,265       31,160      83,425    0.51
FLOOR FINISHES            49,135      232,791     281,926    1.71
CEILING FINISHES          29,875       56,175      86,050    0.52
                       ----------   ----------  ----------   -----
                        1,487,219    2,533,393   4,020,612   24.43
SPECIALTIES & EQUIPMENT

VERTICAL TRANSPORTATION    72,000      108,000     180,000    1.09
SPECIALTIES               17,693      110,103     127,796    0.78
FIXED EQUIPMENT                0      911,245     911,245    5.54
GENERAL CONDITIONS             0      480,000     480,000    2.92
                          ------    ----------  ----------   -----
                          89,693    1,609,348   1,699,041   10.32
HVAC

H.V.A.C.                 477,340      477,340     954,680    5.80
                         -------      -------     -------    -----
                         477,340      477,340     954,680    5.80
PLUMBING

PLUMBING                 325,908      276,528     602,436    3.66
                         -------      -------     -------    -----
                         325,908      276,528     602,436    3.66
```

FIGURE H-3

```
FEBRUARY 20, 1974                                              PAGE   2
                      GROUP-SYSTEM REPORT
-----------------------------------------------------------------------
DESCRIPTION                LABOR      MATERIAL      TOTAL    SQ FT
-----------------------------------------------------------------------

ELECTRICAL

ELECTRICAL               516,844      387,120     903,964    5.49
                         -------      -------     -------    -----
                         516,844      387,120     903,964    5.49

                       ----------   ----------  ----------   -----
CONSTRUCTION TOTAL      2,906,591    5,291,186   8,197,777   49.80
```

FIGURE H-4

DESCRIPTION	QUANTITY	UNIT	LABOR	MATERIAL	TOTAL
ROADS, WALKS & PARKING					
CONCRETE WALKS	6200	SQ FT	6,448	3,596	10,044
CONCRETE PAVING-TO 10 MSF	444	SQ YD	1,871	2,893	4,764
CONCRETE CURBS	400	LN FT	1,268	968	2,236
			-----	-----	------
			9,587	7,457	17,044
FOUNDATIONS					
PIT & TRENCH EXCAV-MACHINE	2711	CU YD	3,199	2,061	5,260
BACKFILL-INCLDS-COMPACT & TA	4200	CU YD	11,131	3,360	14,491
DEWATERING-SET UP & REMOVE	1	ITEM	0	10,000	10,000
PER. DRAINS-PIPE & GRAN. FIL	1360	LN FT	3,563	5,100	8,663
COLUMN FTGS-EXCAV-FMS-RF-CON	1450	CU YD	102,160	75,384	177,544
WALL FTGS-EXCAV-FMS-RF-CONC.	203	CU YD	18,888	12,186	31,074
WATERPROOFING	4080	SQ FT	1,795	694	2,489
INSULATION	4080	SQ FT	571	571	1,142
WALLS-EXCAV-FMS-RF-CONC.	475	CU YD	40,375	32,300	72,675
SHEETING & LAGING	2400	SQ FT	0	3,600	3,600
			--------	--------	--------
			181,682	145,256	326,938
FLOORS ON GRADE					
S.O.G.-FILL-MESH-CONC-FMS-JT	46780	SQ FT	48,183	32,746	80,929
WATERPROOFING	46240	SQ FT	12,022	9,248	21,270
			------	------	-------
			60,205	41,994	102,199
SUPERSTRUCTURE					
ORNAMENTAL METAL COMPLETE	1	LP SM	9,640	11,800	21,440
COLUMNS-FORMS-REINF-CONCRETE	158	CU YD	31,620	18,972	50,592
FLAT SLABS-FORMS-REINF-CONCR	3019	CU YD	452,820	301,880	754,700
CONCRETE STAIRS	5100	SQ FT	49,980	29,070	79,050
			-------	-------	-------
			544,060	361,722	905,782

FIGURE H-5

SYSTEM-COMPONENT REPORT LABOR/MATERIAL

DESCRIPTION	QUANTITY	UNIT	LABOR	MATERIAL	TOTAL
ROOFING					
SKYLIGHT	800	SQ FT	9,600	8,000	17,600
ROOFING-COMPLETE	53100	SQ FT	55,755	42,480	98,235
			------	------	-------
			65,355	50,480	115,835
EXTERIOR WALLS					
ENTRY DOORS	1	EACH	300	1,900	2,200
20'X15' ROLL-UP DOOR	1	EACH	1,125	1,985	3,110
GLASS COURT WALLS	9757	SQ FT	68,299	126,841	195,140
PRECAST PANELS-PLAIN	82933	SQ FT	165,866	1,086,422	1,252,288
ALUMINUM WINDOWS	10300	SQ FT	31,106	42,539	73,645
H.M.-FRAME-DOOR-HARDWARE	12	EACH	1,045	3,000	4,045
			-------	----------	----------
			267,741	1,262,687	1,530,428
PARTITIONS					
5' DEMOUNTABLE PARTITION	1500	LN FT	8,505	47,100	55,605
FINISHED CARPENTRY	1	LP SM	0	170,000	170,000
CONCRETE BLOCK	89340	SQ FT	120,609	41,990	162,599
DRYWALL & METAL STUDS	70720	SQ FT	85,571	28,288	113,859
H.M.-FRAME-DOOR-HARDWARE	255	EACH	22,216	63,750	85,966
			-------	-------	-------
			236,901	351,128	588,029
WALL FINISHES					
PAINTING	89000	SQ FT	25,560	15,900	41,460
CERAMIC TILE	15260	SQ FT	26,705	15,260	41,965
			------	------	------
			52,265	31,160	83,425
FLOOR FINISHES					
RESILIENT FLOORING	27910	SQ FT	8,373	15,351	23,724
CARPET	107227	SQ FT	26,807	176,925	203,732

FIGURE H-6

SYSTEM-COMPONENT REPORT LABOR/MATERIAL

DESCRIPTION	QUANTITY	UNIT	LABOR	MATERIAL	TOTAL
FLOOR FINISHES	(CONT.)				
CERAMIC TILE	4300	SQ FT	7,955	4,515	12,470
PAINTING	20000	SQ FT	6,000	36,000	42,000
			49,135	232,791	281,926
CEILING FINISHES					
METAL PAN CEILING	7500	SQ FT	4,875	11,175	16,050
MECHANICAL ENVELOPE	1	LP SM	25,000	45,000	70,000
			29,875	56,175	86,050
VERTICAL TRANSPORTATION					
PASSENGER ELEVATOR-LOW SPEED	3	ITEM	72,000	108,000	180,000
			72,000	108,000	180,000
SPECIALTIES					
FIRE FIGHTING EQUIPMENT	1	LP SM	4,500	16,800	21,300
TOILET ACCESSORIES	50	EACH	2,544	5,250	7,794
TOILET PARTITION	110	EACH	4,199	15,400	19,599
FOLDING PARTITIONS	3350	SQ FT	5,528	14,238	19,766
LOCKERS	80	EACH	922	3,520	4,442
CHALK/TACKBOARD	677	ITEM	0	50,775	50,775
MOVIE SCREEN	38	ITEM	0	3,800	3,800
MIRRORS	10	ITEM	0	320	320
			17,693	110,103	127,796
FIXED EQUIPMENT					
OBSERVATORY 18' D	1	EACH	0	15,000	15,000
MISC GROUP II EQUIP	1	LP SM	0	863,845	863,845
GROUP 1 MISC	1	LP SM	0	30,000	30,000
CRANES	2	ITEM	0	800	800

FIGURE H-7

SYSTEM-COMPONENT REPORT LABOR/MATERIAL

DESCRIPTION	QUANTITY UNIT	LABOR	MATERIAL	TOTAL
FIXED EQUIPMENT	(CONT.)			
HOISTS	2 ITEM	0	1,600	1,600
		----	-------	-------
GENERAL CONDITIONS		0	911,245	911,245
GENERAL CONDITIONS	1 LP SM	0	480,000	480,000
		----	-------	-------
H.V.A.C.		0	480,000	480,000
HVAC COMPLETE	164600 SQ FT	477,340	477,340	954,680
		-------	-------	-------
PLUMBING		477,340	477,340	954,680
PLUMBING COMPLETE	164600 SQ FT	325,908	276,528	602,436
		-------	-------	-------
ELECTRICAL		325,908	276,528	602,436
125 KW EMERGENCY GENERATOR	1 EACH	0	25,000	25,000
ELECTRIC COMPLETE	164600 SQ FT	516,844	362,120	878,964
		-------	-------	-------
		516,844	387,120	903,964
		---------	---------	---------
CONSTRUCTION TOTAL		2,906,591	5,291,186	8,197,777

FIGURE H-8

DESCRIPTION	TOTAL COST	QUANTITY	UNIT	$/UNIT
ROADS, WALKS & PARKING				
CONCRETE WALKS	10,044	6,200	SQ FT	1.62
CONCRETE PAVING-TO 10 MSF	4,764	444	SQ YD	10.73
CONCRETE CURBS	2,236	400	LN FT	5.59
	17,044			
FOUNDATIONS				
PIT & TRENCH EXCAV-MACHINE	5,260	2,711	CU YD	1.94
BACKFILL-INCLDS-COMPACT & TAMP	14,491	4,200	CU YD	3.45
DEWATERING-SET UP & REMOVE	10,000	1	ITEM	10,000.00
PER. DRAINS-PIPE & GRAN. FILL	8,663	1,360	LN FT	6.37
COLUMN FTGS-EXCAV-FMS-RF-CONC.	177,544	1,450	CU YD	122.44
WALL FTGS-EXCAV-FMS-RF-CONC.	31,074	203	CU YD	153.07
WATERPROOFING	2,489	4,080	SQ FT	0.61
INSULATION	1,142	4,080	SQ FT	0.28
WALLS-EXCAV-FMS-RF-CONC.	72,675	475	CU YD	153.00
SHEETING & LAGING	3,600	2,400	SQ FT	1.50
	326,938			
FLOORS ON GRADE				
S.O.G.-FILL-MESH-CONC-FMS-JTS-FN	80,929	46,780	SQ FT	1.73
WATERPROOFING	21,270	46,240	SQ FT	0.46
	102,199			
SUPERSTRUCTURE				
ORNAMENTAL METAL COMPLETE	21,440	1	LP SM	21,440.00
COLUMNS-FORMS-REINF-CONCRETE	50,592	158	CU YD	320.20
FLAT SLABS-FORMS-REINF-CONCRETE	754,700	3,019	CU YD	249.98
CONCRETE STAIRS	79,050	5,100	SQ FT	15.50
	905,782			

FIGURE H-9

SYSTEM-COMPONENT REPORT UNITS

DESCRIPTION	TOTAL COST	QUANTITY	UNIT	$/UNIT
ROOFING				
SKYLIGHT	17,500	800	SQ FT	22.00
ROOFING-COMPLETE	98,235	53,100	SQ FT	1.85
	115,835			
EXTERIOR WALLS				
ENTRY DOORS	2,200	1	EACH	2,200.00
20'X15' ROLL-UP DOOR	3,110	1	EACH	3,110.00
GLASS COURT WALLS	195,140	9,757	SQ FT	20.00
PRECAST PANELS-PLAIN	1,252,288	82,933	SQ FT	15.10
ALUMINUM WINDOWS	73,645	10,300	SQ FT	7.15
H.M.-FRAME-DOOR-HARDWARE	4,045	12	EACH	337.08
	1,530,428			
PARTITIONS				
5' DEMOUNTABLE PARTITION	55,605	1,500	LN FT	37.07
FINISHED CARPENTRY	170,000	1	LP SM	170,000.00
CONCRETE FLOOR	162,599	89,340	SQ FT	1.82
DRYWALL & METAL STUDS	113,859	70,720	SQ FT	1.61
H.M.-FRAME-DOOR-HARDWARE	85,966	255	EACH	337.12
	588,029			
WALL FINISHES				
PAINTING	41,460	89,000	SQ FT	0.47
CERAMIC TILE	41,965	15,260	SQ FT	2.75
	83,425			
FLOOR FINISHES				
RESILIENT FLOORING	23,724	27,910	SQ FT	0.85
CARPET	203,732	107,227	SQ FT	1.90

FIGURE H-10

SYSTEM-COMPONENT REPORT UNITS

DESCRIPTION	TOTAL COST	QUANTITY	UNIT	$/UNIT
FLOOR FINISHES (CONT.)				
CERAMIC TILE	12,470	4,300	SQ FT	2.90
PAINTING	42,000	20,000	SQ FT	2.10

	281,926			
CEILING FINISHES				
METAL PAN CEILING	16,050	7,500	SQ FT	2.14
MECHANICAL ENVELOPE	70,000	1	LP SM	70,000.00

	86,050			
VERTICAL TRANSPORTATION				
PASSENGER ELEVATOR-LOW SPEED	180,000	3	ITEM	60,000.00

	180,000			
SPECIALTIES				
FIRE FIGHTING EQUIPMENT	21,300	1	LP SM	21,300.00
TOILET ACCESSORIES	7,794	50	EACH	155.88
TOILET PARTITION	19,599	110	EACH	178.17
FOLDING PARTITIONS	19,766	3,350	SQ FT	5.90
LOCKERS	4,442	80	EACH	55.52
CHALK/TACKBOARD	50,775	677	ITEM	75.00
MOVIE SCREEN	3,800	38	ITEM	100.00
MIRRORS	320	10	ITEM	32.00

	127,796			
FIXED EQUIPMENT				
OBSERVATORY 18' D	15,000	1	EACH	15,000.00
MISC GROUP II EQUIP	863,845	1	LP SM	863,844.96
GROUP 1 MISC	30,000	1	LP SM	30,000.00
CRANES	800	2	ITEM	400.00

FIGURE H-11

SYSTEM-COMPONENT REPORT UNITS

DESCRIPTION	TOTAL COST	QUANTITY	UNIT	$/UNIT
FIXED EQUIPMENT (CONT.)				
HOISTS	1,600	2	ITEM	800.00

	911,245			
GENERAL CONDITIONS				
GENERAL CONDITIONS	480,000	1	LP SM	480,000.00

	480,000			
H.V.A.C.				
HVAC COMPLETE	954,680	164,600	SQ FT	5.80

	954,680			
PLUMBING				
PLUMBING COMPLETE	602,436	164,600	SQ FT	3.66

	602,436			
ELECTRICAL				
125 KW EMERGENCY GENERATOR	25,000	1	EACH	25,000.00
ELECTRIC COMPLETE	878,964	164,600	SQ FT	5.34

	903,964			

CONSTRUCTION TOTAL	8,197,777			

FIGURE H-12

10
SCHEMATIC DESIGN

Sometimes referred to as sketch design, the schematic design phase is the initial phase of the actual design work accomplished by the design team. At this point in time, the design concepts generally would already have been developed during the budgetary and programming development, as shown in Table 10-1, and a site plan developed as well.

It is possible, and preferable, that the design team may have participated in the programming phase, and more probably in development of the site plan. If so, the schematic design can be implemented more rapidly.

Since the precedent budgetary, programming, and site-planning work will strongly influence the schematic design, the initial schematic design effort is oriented toward an evaluation of that work. If the prior work has been very well organized and thoroughly executed, the design team will have a strong base upon which to build a design. If, on the other hand, the work has been informal and somewhat nebulous, the design team must proceed cautiously into the schematic phase. In the less well-defined base, the design team must compensate and in effect perform the precedent work itself. The result is an initial stage which involves the development of parameters.

There is no single set of guidelines which cleanly defines this interface between phases. In the case study of the university hospital, a good program had been developed and the schematic design phase had actually been completed when a value-analysis study indicated that the results were not acceptable in financial terms to the university and therefore the entire program was dropped. This is an extreme result of value analysis, and that particular analysis should actually have been made at an earlier stage in the project.

Table 10-1
Design Concepts during Budgetary and Program Development

1. Functional program

a. Mission and specific objectives of project
b. Response to corporate objectives
c. Long-range program envisioned for project
d. Initial program to be undertaken
e. Specific services to be provided
f. Other activities to be accommodated
g. Organizational structure and staffing pattern
h. Proposed operational policies and procedures
i. Anticipated utilization

2. Architectural program

a. Identification and description of project
b. Statement of purpose and objectives
c. Summary of background information
d. Anticipated future developments
e. Site and building requirements
f. Outline of services and functions
g. Schedule of space allocation and equipment
h. Functional data and specific requirements
i. Outline of proposed organization and staffing
j. Summary of proposed operational policy and procedures
k. List of fixed and movable equipment
l. Tentative construction budget

3. Financial program

a. Cost estimate of site and utilities and landscaping
b. Construction cost estimate
c. Estimate of professional fees and expenses
d. Equipment cost estimate
e. Reserve for move-in costs
f. Allowance for overhead and administrative expenses
g. Allocation for project promotion and fund raising
h. Interest and other loan charges during construction
i. Contingency reserve against underestimates and unanticipated costs
j. Tentative project budget
k. Preliminary financial arrangements

4. Site plan

a. Preliminary investigation of suitable sites
b. Site selection and acquisition
c. Survey of plot boundaries and elevations
d. Topographical and geological analysis
e. Study of traffic, access and egress
f. Analysis of zoning regulations
g. Study of potential building volume and land coverage
h. Analysis of climatic exposure, environment and view
i. Preliminary study of approaches, landscaping and parking
j. Master plan of site development

The schematic design phase offers the best single opportunity for the design team to apply value analysis. (In the prior phases, the design team did not actually exist, and the designers' contribution as part of the programming group would be principally as design advisors.) The schematic design is the stage at which the design team should be most receptive to creative value-analysis sessions.

Looking forward, the schematic design should include:

1. Analysis of functions, services, and planning data

2. Development of diagrams of functional relationships and flow of services

3. Blockout of areas by departments, divisions, or functional elements

4. Schematic arrangement of the plan elements

5. Adaptation of schematic layout to site plan and/or site master plan

6. Schematic traffic flow of exterior pedestrian and vehicular traffic

7. Interior traffic flow, including entrances, exits, corridors, stairs, and elevators

8. Development of a building plan, usually on a scale of $1/16$ in. = 1 ft or $1/8$ in. = 1 ft

9. Preparation of major elevation views and some sections

10. Energy analysis, including selection of primary HVAC solution

11. Selection of structural system

12. Specification of major utilities requirements

The schematic design is principally architectural in nature. It is a conversion of the program requirements into an architectural solution which best suits the site. Unless the site presents unusual problems, the input of the engineering disciplines is applied after arriving at the basic architectural solution. For an unusual site, the architect may call upon the structural designer to affirm the feasibility of certain solutions. In projects involving particularly heavy heating or cooling loads, a mechanical engineer may have to furnish an input in terms of size of cooling towers or other special equipment which would directly impact the architectural solution. In isolated projects, the electrical designer, as well as the mechanical, would have an input on utilities connections or energy sources. These also would have an architectural impact, particularly if a power plant were required.

ARCHITECTURAL

The initial stage of the development of space to meet the requirements of the owner is a review of the precedent material. Value analysis at this point in the beginning of the schematic design can be accomplished, as illustrated in the case history on the General Services Administration Social Security payment centers program. The Pareto relationship should be used, and the value analysis should be based on making those large decisions which influence the greatest cost rather than on a room-by-room procedure.

In developing the scheme, the architect also assembles from personal experience and available sources other parameters of design. For example, there are the legal parameters, such as building height. The architect has to know, or find out, that in Washington, D.C. the planning commission by tradition has not approved buildings taller than the capitol, or that in Philadelphia the planning commission is very reluctant to approve high-rise structures with an elevation higher than the cap on the statue of William Penn on top of City Hall. In buildings in aircraft flight paths, the ultimate height must be approved by the Federal Aviation Administration.

In addition to setting of height, the zoning or building code may require that only a certain percentage of the site may be filled by the structure. Also, the structure may occupy one percentage at ground level, but setbacks may be required as the building increases in elevation. The zoning code also often requires a certain number of parking spaces per square foot of building or site.

Basic information in regard to the site is important in terms of the ease of providing basement and subbasement spaces. If excavation is easy and the materials stable, with little or no ground water, a ready solution may be available for garage and utility spaces. The same solution could be very expensive on a site with rock or high ground water.

There are many experience guidelines which can be used in the location of space within a facility, such as the rental value of commercial space located on street level as compared with other levels and locations. The program may direct certain locations or functions. There may also be regulations which preclude certain solutions. For instance, the Bureau of the Budget of New York City will not approve parking structures, a policy probably based on pure economics of site parking at $1,000 to $2,000 per vehicle versus $5,000 to $7,000 for a structure.

Comparisons, however, must be made on the total space solution and not just on the parts. For instance, in accepting a lower cost per unit of parking, the designer could preclude a higher value-per-dollar solution for other functions of a building. In evaluating above-grade parking structure versus excavation in rock, the total value analysis must also consider the added cost of structure for the higher levels above the parking (Fig. 10-1).

In a high-rise building, there is a certain overhead cost per floor which must be absorbed, particularly if the floor is less than optimum size. Maximum value would be achieved per floor or bay of a high rise when the net usable floor space is the maximum which can be supported by a single stairwell, one pair of restrooms per floor, and a single bank of elevators. When the net usable space is less than that figure, the overhead cost per floor is too expensive.

In performing value analysis of space solutions, each feasible space arrangement should be costed out, generally on a square-foot basis. The worth cost should be the lowest cost which could meet the program requirements within building regulations. There is no viable purpose in comparing an urban high-rise situation with a suburban one-story spread-out structure if the latter is not permissible in the area. However, if additional land could be acquired, a study considering total cost including land could be the worth basis.

A reasonable cost comparison of the various alternatives can be achieved by the use of square-foot cost factors based on comparable buildings in the same geographical area. It is preferable that these

VALUE ENGINEERING FUNCTIONAL ANALYSIS WORKSHEET

PROJECT Garage

ITEM Comparison

BASIC FUNCTION Park Cars

NAME

TEL. NO.

DATE

QUANTITY	UNIT	ELEMENT DESCRIPTION	FUNCTION VERB	NOUN	KIND	EXPLANATION	WORTH	COST
460	cars	Steel structure garage – actual	Park	Cars	P			$913,000
460	"	Surface parking				Av/car $1,000	$460,000	
460	"	Garage structure above grade (typical)				Av/car $2,785	$1,281,000	
460	"	Garage structure above grade (typical)				400 sq ft/car @ $25/sq ft	$1,000,000	

FIGURE 10-1
Functional analysis of alternative parking facilities.

costs be escalated to the time of projected construction so that the figures not only are comparable on a relative basis but also represent reasonable figures if taken out of context.

In performing value analysis of a spatial solution, there is generally no single correct answer. Accordingly, a substantial number of solutions may have to be costed in order to arrive at the best value solution. To accomplish this, a computerized cost model is particularly effective, such as the AMIS cost model discussed in Case Study G. In fact, the AMIS model had its basis for development in that type of use. The Health and Mental Hygiene Facilities Improvement Corporation utilized this approach on a number of projects. For the Bronx State School, AMIS developed the computerized system of design analysis which automatically costed out the most economical design solution conforming with the geometry function and character of the facility. On that project, which was programmed at $30 million, 10 different schematic solutions were costed within 24 hours.

A design team implementing a schematic design for an expansion program for a New York City community college achieved a substantial improvement in the cost-worth ratio by dramatically reducing the number of individual buildings, thus reducing the environmental skin. The 11 programmed buildings were reduced to 6 basic structures.

VALUE ENGINEERING FUNCTIONAL ANALYSIS WORKSHEET

PROJECT __College - academic building__   NAME_____

ITEM ____Spatial mix____   TEL. NO._____

BASIC FUNCTION____Provide Classrooms____   DATE_____

QUAN-TITY	UNIT	ELEMENT DESCRIPTION	FUNCTION VERB	NOUN	KIND	EXPLANATION	WORTH	COST
60	Sqft	Classrooms - faculty space & internal circulation	Provide	Space	P	2 wings each floor		
6	Sqft	Toilet rooms - core	Support	Habita-tion	P	Two per floor		
4	Sqft	Stair circulation	Provide	Access	P	Concrete - fireproof		
			Provide	Exits	P			
5	Sqft	Mechanical spaces	Support	Habita-tion	P	Chillers, fans, switchgear		
								$3,975,000
		Cost/gross sqft = $53 /net sqft = $67 /net academic sqft = $80						

FIGURE 10-2
Functional analysis of the spatial mix for a college academic building.

VALUE ENGINEERING FUNCTIONAL ANALYSIS WORKSHEET

PROJECT Development complex

ITEM Spatial mix

BASIC FUNCTION Generate Revenue

NAME _____

TEL. NO. _____

DATE _____

QUAN TITY	UNIT	ELEMENT DESCRIPTION	FUNCTION VERB	NOUN	KIND	Annual Net Revenue (1,000s)	$/sqft EXPLANATION	Est.WORTH $1,000s	Est. COST
484	Sqft	Office tower	Generate	Rentals	P	2,364	4.88	17,177	$22,500
140	Sqft	Apartments	Generate	Rentals	P	382	2.73	3,528	4,272
100	Sqft	Motor inn	Generate	Income	P	675	6.75	4,141	4,940
			Support	Offices	S				
150	Sqft	Retail dept. store	Generate	Income	P	487	3.25	3,106	3,923
40	Sqft	Retail specialty shops	Generate	Income	P	170	4.25	887	1,104
			Create	Atmos-phere	S				
5.5	Sqft	Theater	Generate	Income	P	180	32.73	163	200
270	Sqft	Parking (850 spaces)	Support	Complex	P	133	1.23	2,918	3,775
			Create	Revenue	S				
100	Sqft	Site development	Support	Complex	P			887	1,526
1290	Sqft								

FIGURE 10-3
Functional analysis of the spatial mix for a development complex.

Figure 10-2 shows the spatial mix for a college academic building. The breakdown is limited to four major categories.

In Fig. 10-3, a multiproject mix is shown for a development complex. The worth and cost estimates are based on square-footage ranges. Worth was taken to be the lower range of figures which could be anticipated, while the estimated is the median. Also shown on the chart are figures for the annual revenue and the annual revenue per square foot. The value in this value analysis is to a large degree the amount of revenue which can be generated on a net basis for the developer. There are detailed analyses which support each of the figures shown (see Case Study E).

STRUCTURAL

The spatial solution mandates the general structural solution. For instance, if the spatial solution can accept 20 × 20 ft modular bays, interim columns can be utilized, which quite often offer lower cost per unit of structure. In some instances, the spatial solution requires future flexibility, dictating long-span beams and utilizing non-load-bearing partitions which can be altered in the future. Long span is also a requirement in structures such as gymnasiums and swimming pools.

The spatial solution may also have vertical restrictions. In high-rise structures with a limited elevation, reduction in the floor-to-floor height can result in direct savings by providing additional floor space per cube of building provided and by reducing the exterior skin required per square foot of usable floor space.

One comparison between a concrete and steel structure indicated that the use of a reinforced flat-plate design in concrete would permit a 6$^{1}/_{2}$ in. floor slab, contrasted with the steel solution which would require a 20 in. floor depth. However, in the structural steel building, certain types of ductwork could be recessed into the steel, making the impact somewhat less.

The Concrete Reinforcing Steel Institute (CRSI) described a $125,000 saving which was accomplished by using concrete rather than steel framing (Ref. 1). The project was a high-rise condominium in Philadelphia—a 20-story structure valued at $3,000,000. The structural engineer costed both steel and reinforced concrete. The two basic systems analyzed were:

1. Steel: plastic design with braced frames and composite beams or joist floors

2. Concrete: flat plate with high-strength grade 60 rebar reinforcement using concrete walls for lateral stability

The flat-plate scheme costed out at $850,000. The structural-steel solution was costed by the owner's own engineers at approximately $975,000.

The CRSI indicated that the key to the savings was the flat-plate design which permitted least floor-to-floor height and which precluded the requirement for additional fireproofing. The fireproofing cost had to be included in the cost of the structural-steel scheme. Important secondary advantages to the concrete were its lower transmission values for heat and noise.

The best value choice between structural systems is never automatic. It varies with the cost of materials, geographical location, and trade practices, as well as the type of structure. In New York City, the revised building code authorized an ultimate-strength design factor for reinforced concrete. In a major building, with a superstructure cost of about $7 to $8 million, the designer decided to shift to reinforced concrete from structural steel to take advantage of the revised code. Initially, the cost advantage was estimated to be approximately $100,000. As the design proceeded, this advantage diminished. The project manager had recommended structural steel because of the advantage in phased construction possible by early ordering of the steel. The construction manager (CM) felt that 6 months to 1 year could be saved by early ordering of the steel. The converse occurred when the City Building Department was reluctant to approve the new design because they had little or no precedent experience and none on a building of this size. The final result was an estimated cost approximately the same as steel, but a delay of at least 1 year in a building valued at over $50 million and a loss in escalation of at least $2 million. Comparative analysis during selection of structural systems must be viewed from all dimensions.

There is no single criterion for analysis of structural systems for buildings which can be used to assure that one structural answer will provide the best value for all situations. In fact, the following examples from one campus in which several buildings were being designed simultaneously illustrate that the solution for one building does not provide the best value for another.

The campus administration offices, computer center, and student activities were to be combined into one building. After the spatial solution was developed, a structural engineer suggested three schemes as the most appropriate. These were as follows:

Scheme 1—Flat plate lightweight	$6.26/sq ft
Scheme 2—Structural steel composite beams	$7.62/sq ft
Scheme 3—One-way normal weight	$6.84/sq ft

The design team selected the flat-plate approach for a number of reasons including, but not solely, least cost. The flat plate also provides the least overall height of the building. The architect indicated that the design approach would be an exposed ceiling, and so the flat-plate ceiling was the best finish for this approach. The flat-plate slab is deep and will accommodate imbedded electrical conduit. Also, the structure involved is a complex one, in terms of exterior shape, and the flat plate will be easier to frame into nonrectangular bays. Similarly, it would be easier to frame breaks in slab that occur away from the columns. The flat plate also requires no fireproofing.

The designer noted certain disadvantages of using flat plate, including: more deflection than other systems, difficulty in making future structural changes, and the necessity for some acoustical treatment. From the architectural viewpoint, the architect indicated that acoustics and color would be picked up by the use of wall-to-wall carpet.

Another designer, specifying requirements for a laboratory building, had a similar set of design criteria. The structural engineer for this building considered four schemes to handle the 20 × 20 ft bays supported by 16-in. circular columns:

Scheme 1—Flat plate normal weight	$5.46/sq ft
Scheme 2—Flat plate lightweight concrete	$5.54/sq ft
Scheme 3—Structural-steel composite beams	$6.91/sq ft
Scheme 4—Structural-steel system	$7.68/sq ft

Again, the flat plate was the least expensive, surprisingly less expensive than the flat-plate lightweight concrete. This approach was selected for the same reasons as the flat plate in the administration–student center building.

In the library building, the bay size was larger (25 × 25 ft), and the structure had to support a 150 lb/sq ft live load for the book stacks. Three structural schemes were compared as follows:

Scheme 1—Two-way dome-slab system	$6.87/sq ft
Scheme 2—One-way pan joist system with girders	$7.02/sq ft
Scheme 3—1½-in. steel deck spanning over 16-in. steel beams supported by composite girders	$6.91/sq ft

Scheme 1 was selected on the basis of appearance and inherent fireproof nature.

In the gymnasium swimming pool area, the structural requirements were for long span over the functional areas. The structural designer considered two schemes as follows:

Scheme 1—Precast-prestressed single tees	$7.30/sq ft
Scheme 2—Structural-steel precast-prestressed concrete plank	$9.05/sq ft

On the basis of lower cost and better resistance to corrosion (over the pool area), the precast-prestressed single tees approach was selected.

One of the secondary advantages of reinforced-concrete structure is its innate fireproof characteristics. In *Architectural Record* (Ref. 2), United States Steel disclosed a new concept in fire protection called flame shielding. This approach eliminates the need to cover exterior surfaces with fireproofing material. The basis of this approach is the inclusion of plain or galvanized sheet steel shielding as a part of major spandrel members. The flame shields deflect flame outward away from girders. The main steel members support the cladding.

New concepts in steel framing have led to a combination of advantages. As noted, one of the concepts of design is the provision of large open interior spaces. Research done by United States Steel has produced a revised design for internal wind bracing, resulting in a pattern which produces optimum framing. One design for a 54-story building was able to cut steel weight to 25 lb/sq ft with no columns located in the perimeter office space. An earlier concept used on the United States Steel building in Pittsburgh avoided fireproofing requirements by providing structural tubes in the exterior of the building through which water circulated to maintain appropriate structural temperatures in the event of a fire.

In Ref. 3 a new concept for high rise is described. This is similar to the concept described in Ref. 2 which is called the tubular design approach. The skin of the building becomes a tube through the use of relatively closely spaced columns. The structure described is actually a hexagonal tube with rigid frames. The designers believe that the steel is reduced by about 30 percent. Andrew Gravino, partner in the design firm, estimated that a conventional rigid frame would have had a structural steel weight of about 26 to 28 lb/sq ft. The weight of the building steel is projected to be about 19.5 lb/sq ft. Further, the design indicates that if the building had been rectangular, that weight could have been reduced to 16 lb/sq ft—a dramatic reduction in the weight and thereby the cost. This approach also eliminates interior bracing because wind load is handled entirely by the exterior columns.

One concern in projecting dramatic savings through value analysis is the assurance that the field work can be installed as contemplated during design. In Ref. 4, the designers selected a steel support for a major ($2.6 million) bridge overpass in Cleveland which utilized V-shaped structural members called delta frames. These frames require extensive shop welding. The engineering office which originally did the design work contemplated that the structure would be easier to erect in the field and that the net saving would be about $50,000 as a result of easier field erection. The actual result was quite the reverse. The bridge was built to DOT and AWS (American Welding Society) standards, as well as Ohio state standards. The state standards appear to have been more stringent than the AWS, and quite a few of the welds were rejected. In fact, only 20 percent were approved and 80 percent required either additional radiographic testing or repairs. Many of the shop welds were rejected as a result of field radiographs, thus requiring more costly field repairs.

Another result of the structural system is the foundation system required to support it. While the foundation is the first segment of a building to be constructed, it is the last to be designed. The structural frame dictates to a large degree the weight which the foundations must carry. Also, during schematic design, design of the foundations must fall within the required parameters of the using agency and the building code. Californians have for years been accustomed to design according to seismic loading. More recently, on the East coast, buildings being designed for the U.S. Army and a number of other major agencies are now required to withstand seismic loads. In addition to the dead load imposed by the structure and the live load required to be carried, the configuration of the structure and its inherent stiffness directly affects the loads which the foundation must carry. Value analysis can be applied to foundation design. Often, designers are quite conservative in the foundation area and design according to their own span of experience and conservatism. There are new approaches available. For instance, Vibroflotation Foundation Company cites a situation in which the designers were able to use their approach for compacting a sand base as a means of saving a substantial amount in foundation costs. Vibroflotation is well known in sandy areas such as Florida. This particular application was in Akron, Ohio. The designers, recognizing a low bearing capacity in the existing soil, specified a 36-in. thick reinforced-concrete mat. Vibroflotation was able to compact the substrata to a depth of 25 ft. The savings in concrete material alone was more than the cost of Vibroflotation.

HEATING, VENTILATING, AND AIR CONDITIONING (HVAC)

Given the spatial solution to the facility requirements, the geographical location, and the specified comfort range for inhabitants in terms of average temperatures and humidity, the mechanical engineer develops the HVAC schematic design. The first step is determination of the heating and cooling loads. The universal medium for delivery of heat (or removal of it in hot weather) is air. Usually, the next step is the selection of heating and cooling energy sources. The remaining major selection is the distribution system between the energy source and the point of delivery.

The HVAC designer must consider the special requirements of the facility, as well as special availabilities. For instance, many cities have available low-pressure exhaust steam from power plants, which provides a very convenient heat source and permits a lower first cost for the heating system. Water-pollution treatment plants produce methane gas as a by-product and may be able to convert this into a heat source to provide steam for the facility. For cooling, an availability of cooling tower water in a process plant could provide an inexpensive form of condensing expanded refrigerant gases. More usually, in hot weather, heat is available in the form of exhaust steam or by-products of refrigeration machines used for cooling. The use of absorption-type refrigeration equipment provides a means for converting this heat into an energy source for cooling.

The location of the building in relation to the sun will provide specific heating and cooling loads as well as a shifting of the load during the course of a day or a season. The HVAC designer must develop a heating/cooling analysis and the system to change the environment to the desired conditions.

When fuels were less expensive, many of the possible solutions were essentially comparable in terms of operating costs, and it was difficult to justify a higher initial installed cost in terms of lower operating costs. Today, the disparity between systems is becoming more and more dramatic.

In *Building Design & Construction,* June 1974 (Ref. 5), a special utilization is described at the Hunt's Point Cooperative Market located in the Bronx, New York. This complex has five buildings on 37 acres. Half of the facility is refrigerated to handle wholesale meat and poultry supplies. A system has been installed to recover heat from the prime refrigerant, which is a calcium chloride brine solution. It is estimated that recovered heat will save more than $100,000 per year by reducing the fuel-oil consumption by more than $200,000 annually. A secondary advantage is reduction of air pollution.

Further, the heat recovery system will permit the installation of a smaller boiler, which will essentially match in savings the cost of the heat-recovery system itself.

The spatial solution has an impact upon the cost of the mechanical HVAC systems. In one instance, the design engineer agreed to avoid installation of fresh air louvers at the midpoint on a high-rise building. The result was oversized utility-line requirements and oversized fans and pumps. The value impact was substantial but not visible outside of the machinery spaces. Many large buildings, high rise or industrialized low rise, arrange for HVAC modules of optimum size. This is accomplished in high-rise structures by placing mechanical floors at interim levels and in low-rise buildings by the core-utility concept.

The principal heating systems are hot air, hot water, high temperature hot water, and electric. Electric heating tends to be feasible principally in climates where heating is required only in a limited portion of the year. Where heating requirements are most severe, hot water or high temperature hot water (HTW) are most efficient and have the lowest cost distribution system. However, the systems cannot be converted for hot weather use unless a chilled-water cooling system is utilized. In temperate zones, wide utilization is made of the hot-air system. Unit heater and unitary heaters are used as distribution devices for the hot-water and HTW systems.

Air conditioning units utilize hot air or chilled water for distribution. The most common systems are:

1. Single duct, high velocity, variable volume with terminal reheat

2. Dual duct, high velocity, variable volume distribution with mixing boxes

3. Single duct, low velocity, constant volume with terminal reheat

The low velocity is the traditional distribution system, with usually the lowest installation cost. However, actual operating costs are usually lower for the variable-volume and high-velocity systems. The proper utilization of zoning can improve the operating characteristics of any of the distribution systems.

The question of operating cost versus initial cost is an equation which varies with the owner and using agency. A developer building for his own account, but intending to sell a building within 3 to 7 years, would be more inclined to be interested only in higher first cost when it has a complete payoff in 3 to 5 years. An owner building for his own account and intending to operate a building over a long term would accept a longer payback, on the order of 10 to 20 years. A speculative builder intending to sell the building at the earliest possible time would attempt to minimize first cost, often in the face of severe operating cost penalties.

Selection of the chilling unit for air conditioning is discussed under the utilities selection, since it is closely related to the electric power system selection. Similarly, the primary source of heat is interrelated with the electric power solution. The HVAC designer analyzes the distribution approaches. These are directly related to the spatial solution and, in turn, have direct impact upon the preliminary spatial design. In the student center, the designer considered three duct-distribution systems:

1. Central system: single duct, low velocity, constant volume with thermal reheat

2. Central system: single duct, high velocity, constant volume with thermal reheat

3. Decentralized: single duct, low velocity, constant volume with thermal reheat

The system evaluation did not include the computer center which had its own special performance requirements, particularly for humidity. A self-contained system was designed for installation in the computer center. For different reasons, the physical education areas were served by their own separate large system.

Over the years, analysis of duct-distribution systems has focused principally upon lowest first cost. However, because of energy awareness, the analysis included a life-cycle or operating-cost factor. The cost comparisons for the three schemes were:

	Initial Capital Cost	Annual Operating Cost
Scheme 1	441,000	60,861
Scheme 2	465,000	69,207
Scheme 3	455,000	55,079

In this particular case, scheme 3, the decentralized single-duct low-velocity system was more compatible with the operation of the other two major decentralized systems for the computer center and the physical education areas. Even though the initial cost was about $14,000 more than scheme 1, the savings in annual operating costs would pay this back in slightly more than 3 years. In the classroom building, another designer considered the following three schemes:

1. Single duct, high velocity, variable volume with terminal reheat

2. Dual duct, high velocity, variable volume

3. Single duct, low velocity, constant volume, terminal reheat

Unitary equipment was not considered because heat-recovery devices for preheating or precooling outside air cannot be utilized with this approach.

For evaluation purposes, the schemes were costed as follows:

	Initial Capital Cost	Annual Operating Cost
Scheme 1	881,000	100,648
Scheme 2	924,000	95,054
Scheme 3	846,000	128,764

The scheme selected was a single-duct high-velocity variable-volume system, even though its operating cost was slightly more than that of a dual-duct system. The annual advantage of a dual-duct system was not sufficient to provide a payback to the owner in a reasonable period of time. However, any substantial rise in electrical or fuel costs could suggest reevaluation of scheme 2. The single-duct system, although offering low initial capital cost was not low enough in cost to afford the higher annual operating costs, which was almost 30 percent above the system selected.

For most of these schemes, supplementary perimeter heating with fin tube would be employed.

ELECTRICAL DESIGN

At the schematic stage of design, the electrical design is generally limited to an overview study of the basic means of providing electrical energy, since detailed electrical design depends upon a definition of the loads to be serviced, which requires spatial relationships and location of mechanical and electrical equipment.

Remote locations produce special design conditions electrically. Before the universal advent of reliable electrical utility power, most installations had their own electrical energy source. In the past 50 years, this trend completely reversed and few had their own power source if commercial power was available. Today, with high energy costs, a return to the total energy concept is gaining momentum.

The usual prime mover for a total energy source is the gas turbine, which is relatively inefficient. Successful application requires utilization of the exhaust heat from the unit in a heat-recovery boiler. The economical evaluation must include effective utilization of the waste heat, in hot weather as well as cold. This tends to link the total energy concept to the use of absorption refrigeration equipment.

The New York State Dormitory Authority, as part of the development of the Richmond College Campus of the City University of New York (CUNY), has developed a selective energy approach. As reported in *Building Design & Construction* (Ref. 6), this approach resulted from a study by Michael Baker, Inc., and uses on-site generation for peak demand periods. The selective energy is provided for only part of the campus requirements, while standard utility power from Consolidated Edison furnishes a background or reserve amount. This maintains the role of the utility and provides on-site generation capability which can also meet emergency requirements.

REFERENCES

1. *Engineering News-Record*, July 12, 1973, p. 56.

2. *Architectural Record*, March 1974, p. 61.

3. *Engineering News-Record,* October 4, 1973, p. 36.
4. *Engineering News-Record,* April 25, 1974, p. 12.
5. *Building Design & Construction,* June 1974, p. 65.
6. *Building Design & Construction,* July 1974, p. 25.

CASE STUDY I
SCHEMATIC DESIGN (LOCATION OF AUDITORIUM: GSA)

One of the value-analysis teams studying the three SSA payment centers had the responsibility to locate the auditorium.

Design criteria prepared by SSA and furnished to the value-analysis team included:

> Space shall be capable of being subdivided into three rooms by use of two folding acoustical partitions. A lobby and reception room shall adjoin the auditorium. The wall between the lobby-reception area shall also be folding acoustical partitions. Ceiling height of the entire area shall be 14 ft. Auditorium shall be free of columns and equipped with specified audiovisual equipment and projection room of approximately 100 sq ft; a stage platform shall be provided. Programmed space for the auditorium shall be approximately 8,000 sq ft. The team addressed five alternative locations identified by the following scheme numbers:

Scheme	Location
1	Ground floor
2	Top floor
3	Midfloor
4	Separate building
5	Separate building with cafeteria

Figure I-1 shows schematically the five location schemes.

Figure I-2 is a diagram of major and subordinate functions of the auditorium.

Figure I-3 (a) through (e) shows the functional analysis of the five location schemes as performed by the team.

Figure 1-4 (a) and (b) is the proposal summary developed by the team.

CONCENTRATED STUDY AREA

FUNCTION OF PROJECT

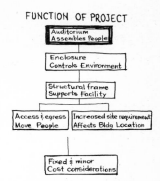

FIGURE I-1
Schematic showing the
five alternatives for the location of the auditorium.

FIGURE I-2
Function diagram for the
auditorium.

VALUE ENGINEERING FUNCTIONAL ANALYSIS WORKSHEET

PROJECT *SSA*

ITEM *Auditorium study*

BASIC FUNCTION *Assemble People*

SCHEME I

QUAN TITY	UNIT	ELEMENT DESCRIPTION	FUNCTION VERB	NOUN	KIND	EXPLANATION	WORTH	Initial COST
1	Sys	Enclosure	Controls	Environment	B	Includes HVAC Lighting fixtug		98,000
1	↓	Structural frame	Supports	Facility	S	Influence of Location on cost		120,000
1		Space dividers	Subdivide	Area	S	Partitions		19,000
1		Sound control	Implement	Accoustics	S	Accoustics		21,000
1		Stage	Elevates	Speakers	S			8,000
1		Training aids	Facilitates	Production	S	Incl. screen proj. room T.V. cable		30,000
1		Access & egress	Move	People	S	Incl. lobby & recept. area & elev.		75,000
1		Seating system	Positions	People	S	Seats		24,000
1		Toilet	Provides	Comfort	S			5,000
1		Kitchen	Provides	Refreshment	S			5,000
1		Site	Affects	Location	S			N.A.
1		Offstage rooms	Supplements	Stage	S	Preparation, Storage, etc.		8,000
12								413,000
						10% gen'l. conditions		41,300
								454,300

I-GROUND FLOOR

FIGURE I-3a
Functional analysis of the ground floor.

PROJECT *SSA*
ITEM *Auditorium study*
BASIC FUNCTION *Assemble People*

SCHEME II

QUANTITY	UNIT	ELEMENT DESCRIPTION	FUNCTION VERB	NOUN	KIND	EXPLANATION	WORTH	Initial COST
1	Sys	Enclosure	Controls	Environment	B	Includes HVAC Lighting fixture		98,000
1	↓	Structural frame	Supports	Facility	S	Influence of location on cost		95,000
1		Space dividers	Subdivide	Area	S	Partitions		19,000
1		Sound control	Implement	Acoustics	S	Accoustics		21,000
1		Stage	Elevates	Speakers	S			8,000
1		Training aids	Facilitates	Production	S	Incl. screen proj. room T.V. cable		30,000
1		Access & egress	Move	People	S	Incl. lobby & recept. area & elev.		195,000
1		Seating system	Positions	People	S	Seats		24,000
1		Toilet	Provides	Comfort	S			7,000
1		Kitchen	Provides	Refreshment	S			5,000
1		Site	Affects	Location	S			N.A.
1		Offstage rooms	Supplements	Stage	S	Preparation, Storage, etc.		8,000
12								510,000
								51,000
						10% gen'l. conditions		561,000

II - TOP FLOOR

FIGURE I-3b
Functional analysis of the top floor.

PROJECT *SSA*
ITEM *Auditorium study*
BASIC FUNCTION *Assemble People*

SCHEME III

QUANTITY	UNIT	ELEMENT DESCRIPTION	FUNCTION VERB	NOUN	KIND	EXPLANATION	WORTH	Initial COST
1	Sys	Enclosure	Controls	Environment	B	Includes HVAC Lighting fixture		98,000
1	↓	Structural frame	Supports	Facility	S	Influence of location on cost		100,000
1		Space dividers	Subdivide	Area	S	Partitions		19,000
1		Sound control	Implement	Acoustics	S	Accoustics		21,000
1		Stage	Elevates	Speakers	S			8,000
1		Training aids	Facilitates	Production	S	Incl. screen proj. room T.V. cable		30,000
1		Access & egress	Move	People	S	Incl. lobby & recept. area & elev.		195,000
1		Seating system	Positions	People	S	Seats		24,000
1		Toilet	Provides	Comfort	S			6,000
1		Kitchen	Provides	Refreshment	S			5,000
1		Site	Affects	Location	S			N.A.
1		Offstage rooms	Supplements	Stage	S	Preparation, Storage, etc.		8,000
12								514,000
								51,400
						10% gen'l. conditions		565,400

III - MID FLOOR

FIGURE I-3c
Functional analysis of the midfloor location.

PROJECT _SSA_
ITEM _Auditorium study_
BASIC FUNCTION _Assemble People_

SCHEME IV

QUANTITY	UNIT	ELEMENT DESCRIPTION	FUNCTION VERB	NOUN	KIND	EXPLANATION	WORTH	Initial COST
1	sys	Enclosure	Controls	Environment	B	Includes HVAC Lighting fixture		131,000
1	↓	Structural frame	Supports	Facility	S	Influence of Location on cost		99,000
1		Space dividers	Subdivide	Area	S	Partitions		19,000
1		Sound control	Implement	Accoustics	S	Accoustics		21,000
1		Stage	Elevates	Speakers	S			8,000
1		Training aids	Facilitates	Production	S	Incl. screen proj. room T.V. cable		30,000
1		Access & egress	Move	People	S	Incl. lobby & recept. area & elev.		65,000
1		Seating system	Positions	People	S	Seats		24,000
1		Toilet	Provides	Comfort	S			5,000
1		Kitchen	Provides	Refreshment	S			5,000
1		Site	Affects	Location	S			90,000
1		Offstage rooms	Supplements	Stage	S	Preparation, Storage, etc.		8,000
12								505,000
								50,500
						10% gen'l. conditions		
								555,500

IV-SEPARATE BDG

FIGURE I-3d
Functional analysis of separate building.

PROJECT _SSA_
ITEM _Auditorium study_
BASIC FUNCTION _Assemble People_

SCHEME V

QUANTITY	UNIT	ELEMENT DESCRIPTION	FUNCTION VERB	NOUN	KIND	EXPLANATION	WORTH	Initial COST
1	sys	Enclosure	Controls	Environment	B	Includes HVAC Lighting fixture		131,000
1	↓	Structural frame	Supports	Facility	S	Influence of Location on cost		99,000
1		Space dividers	Subdivide	Area	S	Partitions		19,000
1		Sound control	Implement	Accoustics	S	Accoustics		21,000
1		Stage	Elevates	Speakers	S			8,000
1		Training aids	Facilitates	Production	S	Incl. screen proj. room T.V. cable		30,000
1		Access & egress	Move	People	S	Incl. lobby & recept. area & elev.		65,000
1		Seating system	Positions	People	S	Seats		24,000
1		Toilet	Provides	Comfort	S			3,000
1		Kitchen	Provides	Refreshment	S			N.A.
1		Site	Affects	Location	S			90,000
1		Offstage rooms	Supplements	Stage	S	Preparation, Storage, etc.		8,000
12								498,000
								49,800
						10% gen'l. conditions		
								547,800

V-SEPARATE BDG WITH CAFETERIA

FIGURE I-3e
Functional analysis of separate building with cafeteria.

VALUE ENGINEERING PROPOSAL SUMMARY		DATE 2/18/72

TO: G.S.A.	FROM: Team #4 value eng. workshop	PHONE NO

ITEM NAME Auditorium		

COMPONENT OF SSA payment center	QUANTITY One	STUDY SPAN: START 2/14/72 COMPLETE 2/18/72

FUNCTION OF ITEM:		ESTIMATED SAVINGS:
VERB Assemble	NOUN People	N.A.

DESCRIPTION OF PRESENT DESIGN

An 8000 S.F. column-free auditorium with 14 ft ceiling height for the assembly of 8000 people with a 2500 sq ft lobby & reception area, all with the capability of being subdivided, and with elevated stage, projection room and training aids.

Purpose: To establish basic function of auditorium, to Assemble People, and as established by study, to determine best location based on design criteria:

> ground floor
> top floor
> mid floor
> separate bldg.
> separate bldg w/cafeteria

To arrive at initial cost comparison, all elements and their functions were listed, described, analyzed, and priced. The ground floor location was priced and used as a base to relate the other four locations.

Intangible costs and maintenance costs were investigated within the limits of the alloted time, but did not reach the end of final dollar assessment.

CONCLUSIONS

Initial cost:

Location of auditorium	Initial cost
Ground floor	$413
Top floor	510
Middle floor	514
Separate bldg.	505
Separate bldg. w/cafeteria	498

The basic factor determining location is the site cost requirements.

RECOMMENDATIONS

Based on lowest initial cost only, it is recommended that the auditorium be located on the ground floor within the building structure.

A separate building in conjunction with the cafeteria should be considered if ultimate site cost and further evaluation of intangible consideration justify.

	Initial cost	incl. site cost
Separate bldg. w/cafeteria	$498	$90
Ground floor location	413	0.
	$ 85 (Represents $\frac{1}{2}$ of 1%	
	of est. total const. cost per bldg.)	

(a)

FIGURE I-4

(a) and (b) Value-engineering proposal summary for auditorium location.

VALUE ENGINEERING PROPOSAL SUMMARY

RECOMMENDATIONS

Intangible benefits inherent in separate bldg. w/cafeteria?
    Free Up bldg. "systems"
Better aesthetics & building design possibilities:
        Security control
        Community use
Separation of major functions from office area:
        Potential for combination w/ground flr. location

FINANCIAL ASPECTS

SUMMARY OF COST OF ORIGINAL DESIGN:

AUDITORIUM LOCATION - INITIAL COSTS

	Gnd flr.	Top flr.	Mid flr.	Sep. bldg.	Sept. bldg. w/cafeteria
Enclosure	$98.	$98	$98	$131	$131
Structural	120.	95.	100.	99.	99.
Access/egress*	75.	195.	195.	65.	65.
Site	0.	0.	0.	90.	90.
Toilets	5.	7.	6.	5.	3.
Kitchen	5.	5.	5.	5.	0.
Fixed costs:	110.	110.	110.	110.	110.
Total costs:	413.	510.	514.	505.	498.
Penalty costs	0.	102.	102.	92.	85.

SUMMARY OF COST OF PROPOSED DESIGN (W/O IMPLEMENTATION COSTS):

        Add 10% genl. cond.

*Includes lobby, reception area & elevators.

        N.A.

SUMMARY OF IMPLEMENTATION COSTS FOR PROPOSED DESIGN:

        N.A.

NET SAVINGS:

        PER UNIT: $ _____

        TOTAL: $ _____

(b)

CASE STUDY J
SCHEMATIC DESIGN (SPATIAL REQUIREMENTS—FILING: GSA)

One of the SSA payment center value teams reviewed the requirements for filing. A major portion of each of the SSA payment centers is assigned to file storage. The team was given the following orientation material.

Design Criteria:
Each of the major operating branches has a requirement for hard copy storage and control area for files, claims, etc., awaiting processing in the operating branches. Each of these major operating branches requires approximately 3,000 to 4,000 square feet of hard copy temporary storage. It is essential that the hard copy temporary storage be in close proximity to the operating personnel in order that they can efficiently conduct their information processing operation. Additionally, the issue of control and responsibility of the file folders is improved when the branch is near its own material.

Standard for Files:
For estimating purposes, each file cabinet requires seven square feet of space. This number includes space for the cabinet with the file drawer opened, file clerk work space and interior flow of people and material. The degree of use for Payment Center files is classified as "very active" and should be laid out with 42" interior aisles between rows of file cabinets.

The design of the building will have significant effect on the layout of files due to the location of the columns, mechanical and electrical rooms, restrooms and primary and secondary corridors. Experience has shown that the layout factor and corridor circulation factor in file areas range 3% to 7% respectively.

To determine the total space required for files, it is necessary to calculate the estimated number of files required by each Payment Center and add a layout factor of 5% and a circulation factor of 10%.

TEAM ACTIVITIES

The initial team activity was a brainstorming session. Some 41 ideas were listed, as shown in Table J-1.

The team determined that each of these payment centers has a requirement of hard copy storage and that all claims documentation and records for change in status must be maintained in file folders. These folders are presently stored in five-drawer file cabinets which have 120 in. of filing per cabinet. The folders weigh 2.1 lb/in., and so each cabinet weighs 176 lb. The cost is approximately $75 per filing inch. A standard five-drawer cabinet requires 7 sq ft of space inclusive of aisles in a typical records maintenance branch of the payment center. Aisles may range from 36 to 60 in., and experience has shown that a corridor circulation factor in file areas ranges from 3 to 7 percent.

Table J-1
Brainstorming Session

1. No storage	23. Eliminate file withdrawal
2. One floor—all storage	24. Cube usage
3. Put in boxes	25. Substitutes for 5-dr. file
4. Use microfilm	26. Library stack height
5. Microfiche	27. Mechanized retrieval
6. Magnetic tape	28. Give file to beneficiary in microfiche form
7. Reduce retention period	
8. Standardize form	29. Give to D.O.'s
9. Send to bank	30. Give to state
10. Contract storage	31. Use mechanical card
11. Increase # of P.C.	32. Supreme Halifax system
12. Disperse files	33. Use sector of bldg. for rotary mechanized file segments
13. Condense form	34. Use sliding track equipment (high density shelving)
14. Use one card for data	
15. Reevaluate operation	35. Compress files
16. Reorganize BSRI	36. Reduce ceiling height
17. Allow use of step ladders	37. Tiered mezzanine
18. Put files in bldg. sandwich	38. Perimeter storage
19. Files part of structure	39. Lower level storage
20. Remote storage	40. Underground storage
21. Pneumatic tubes	41. Structural files up through center of bldg.
22. Conveyors	

To determine the total space required for files at any of the payment centers, the team referred to the *Statement of requirements for trust fund construction of Payment Centers,* SSA Publication No. 14-71 (2-71).

The team had available the Chicago payment center requirements to use in developing details. Projected to 1975, the Chicago center will house 6,225,382 folders requiring 1,431,837 filing inches with a weight of 3 million lb.

The team considered six systems approaches. The results are described in Table J-1. Figure J-1 is the team summary report. Figure J-2 shows some of the actual charts used by the team in presenting their value analysis.

DISCUSSION

Although the recommendation described above is considered to be feasible, this recommendation is subject to "Program" definition for the future by SSA/BRSI.

VALUE ENGINEERING PROPOSAL SUMMARY	DATE 2/18/72

TO: General Services Admin.	FROM: V.E. Team #1 At L.A. Daly, Co. Washington, D.C.	PHONE NO

ITEM NAME

Filing Systems

COMPONENT OF SSA Payment Center	QUANTITY See Attached Material	STUDY SPAN: START 2/14/72 COMPLETE 2/18/72

FUNCTION OF ITEM:	ESTIMATED SAVINGS:
VERB | NOUN Provide | Storage	To Be Verified by Executive Architect

DESCRIPTION OF PRESENT DESIGN

Each of the payment centers has a requirement for hard copy storage. All claims documentation and records of change in status are maintained in file folders. The folders are stored in standard 5-drawer file cabinets. There are 120 filing inches/cabinet. The folders weigh 2.1 pounds/inch. The cabinet weighs 176 pounds at a cost of approximately $.75 per filing inch. A standard 5-drawer cabinet requires 7 square feet of space inclusive of aisles in a typical Records Maintenance Branch of the Payment Center. Aisles may range from 36" to 60". Experience has shown that a layout factor and corridor circulation factor in files areas range 3% to 7% respectively. To determine the total space required for files in any of the payment centers, refer to the "Statement of Requirements" for Trust Fund Construction of Payment Centers, SSA publication No. 14-71 (2-71). For the purpose of this study, the Chicago Payment Center requirements were included in developing details. Projected to 1975 this P.C. will house 6,225,382 folders, requiring 1,431,837 filing inches with a weight rounded to 3,000,000 pounds. Refer to attached details for analysis and cost data. EXHIBIT I IS A SUMMARY OF THE ANALYSIS.

CONCLUSIONS

The team concluded that a self-supported system would provide the greatest flexibility to SSA giving consideration to minimizing the requirement to withdraw the folder for changes in status; or a process oriented work flow. There are overriding considerations to be given relative to equipment costs and design which are unknown to the team. These considerations should be fully explored with SSA/BRSI from the standpoint of feasibility.

RECOMMENDATIONS

1. Construct the building with a self-supported storage capability providing SSA with unlimited flexibility relative to file density, present organization and future organization. This recommendation provides the greates potential saving in storage space and initial cost but may be influenced by future equipment costs. It provides for maximum utilization of space for files and office area. The concept for this recommendation requires a structure not supported by floors—typical floor load for general purposes space will be 80 lbs./sq.ft. This analysis is item Number 4 in the attached details.

"System Building" inherently provides us with the capability of accommodating any self-supporting folder storage system with a minimum of difficulty.

FIGURE J-1
Value-engineering team summary report on filing systems.

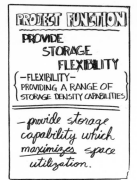

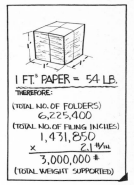

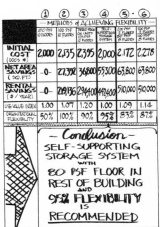

FIGURE J-2
Five cartoon-charts presenting summary of projected SSA filing system costs.

In the event that SSA/BRSI deems this recommendation to be impractical, consideration may then be directed to items 2 and 5 of the summary of results in Table J-2.

Item 1 establishes a floor load of 100 lb/sq ft and is receptive to a mini-payment center, process organization, and increased file density using tiered stack-on units up to 9 ft high with 42-in. aisles.

Item 5 establishes a floor load of 200 lb/sq ft for two floors and 80 lb/sq ft for all other floors. This recommended item fixes the location of files storage on the 200 lb/sq ft floor, thereby prohibiting the distribution of files in a mini-payment center arrangement. It nevertheless provides the greater flexibility for high-density filing equipment.

Table J-2
Summary of Results of Filing-System Study

System	1 5-drawer cabinets	2 7-tier stacks stack-on	3 9-tier stacks stack-on	4 Self-supporting	5 Floor-supported	6 Floor-supported
Distributed load, lb/sq ft	80	100	125	80	200 (2 flrs.)	200 (2 flrs.)
Floors	10	10	10	10	10	10
Construction cost	$2,000,000	$2,135,000	$2,395,000	$2,000,000	$2,172,000	$2,275,000
Area sq ft	91,306	63,914	54,500	36,000	27,500	27,500
Net area saving	0	27,392	36,800	55,300	63,800	63,800
Rental savings	0	219,136	294,400	442,400	510,000	510,000
Index of value	1	1.07	1.2	1.0	1.09	1.14
Deferred construction	No	Yes ($10,000,000)	Yes ($10,000,000)	Yes ($10,000,000)	Yes ($10,000,000)	Yes ($10,000,000)
Flexibility	50%	100%	90%	95%	83%	87%

11
PRELIMINARY DESIGN PHASE

The preliminary design phase is sometimes called design development, which perhaps more accurately describes this stage's purpose
and activities. During the schematic or conceptual design phase, the
basic design concept was defined and approved. The level of definition was sufficient for evaluation and approval but not sufficient in
detail for defining the bounds of a contract.

The purpose of the preliminary design phase is the testing and
development of the basic concept. At the conclusion of this phase, the
working documents can be initiated with a much greater degree of
confidence.

The basic stages in the preliminary design include:

1. Unit studies of the plan elements, determining the feasibility of the
   concept and the availability of acceptable components to meet this
   concept

2. Modular analysis of typical spaces to determine constructability and
   cost practicality

3. Preparation of architectural plans, usually on a scale of $1/8$ in. $= 1$ ft,
   in order to lay out more accurately the basic spatial solution

4. Development of a tentative scheme of mechanical systems and tentative scheme of electrical systems

5. Development of personnel and equipment layouts for functional areas

6. Development of the exterior design of the structure

7. Preparation of a general outline of materials and finishes

8. Takeoff of building areas and preparation of cost estimates

In a sense, this stage of design is a reiteration of the initial stage. The spatial solution proceeds, but more rapidly. Usually, early in the preliminary design, floor plan layouts are made available to the mechanical and electrical engineering disciplines so that they can begin layout of their distribution systems. During this layout, problems will be encountered and fed back to the architectural designers for resolution.

SPATIAL SOLUTION

The preliminary design development brings the structural engineer and architectural team into focus on the solution for the layout approved. In a description from *Engineering News-Record* (Ref. 1), the preliminary structural design had been completed when preliminary estimates were made. The project was a $7 million domed convention center in Indonesia being built by a developer from Singapore. Since it was a development project, the contractors had been selected and a representative of Stolte, Inc., of California estimated both money and time. The representative's indication was that the structure could not be completed on schedule as designed. One of the main features was 16 V-shaped columns of steel covered with concrete. The contractor recommended that these be replaced with 16 Y-shaped columns of reinforced concrete which could be erected more rapidly. Usually, structural steel is more rapidly erected than reinforced concrete, but the V-shaped columns were steel completely encased in concrete, and so a savings of about 1 month was gained by essentially removing the first step in the two-stage process. The concrete columns were also about $25,000 less expensive.

The contractor cooperated with the designer in developing methods for prestressing the concourse deck and dome retaining ring, thus saving both time and approximately $50,000.

The greatest value saving occurred in getting rid of a time-delaying problem. The contractor had estimated that the installation of a partial basement would impose a tremendous time delay because the ground floor slab had to be able to carry the weight of the cranes for erecting the dome. A serial installation of basement and then reinforcement of the floor slab would have been very time-consuming. The contractor got the designers to consider the spatial solution and the partial basement was eliminated by substituting available space behind the seating. This elimination was estimated to save $300,000, and removed 3 months from the original construction schedule.

UNIT STUDY AND
MODULAR ANALYSIS

The basic spatial solution was developed and approved in the schematic design phase. In most buildings, and in particular in offices, schools, and hospitals, the solution must evolve into a three-dimensional treatment in the preliminary design. The schematic design tends to look at things in plan view. However, as indicated in Fig. 11-1, the floor-ceiling sandwich must be considered. The utility delivery resolution requires provision of space for electrical, HVAC, plumbing and communications distribution. A modular solution to this standard problem has been a continuing stumbling block in the evolution of modular building systems.

In the postwar era, one of the most innovational and economical solutions to this question has been the use of the hung acoustical ceiling with a fire rating where required. The approach offers flexibility and meets performance requirements in most ways. The cost of the hung ceiling at $1 to $2/sq ft installed plus its ease of installation is difficult to improve upon.

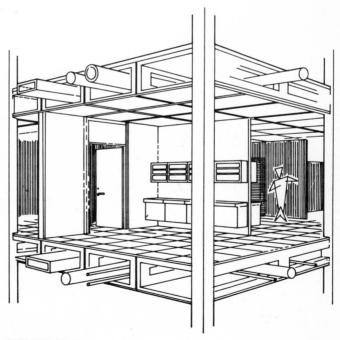

FIGURE 11-1
The floor-ceiling sandwich. (From a National Bureau of Standards study for PBS/GSA, used in *Contractor's Management Handbook.*)

One of the problems not resolved by the overhead hung ceiling is the desk-side availability of power and communications cabling for office structures. Office layouts are in a constant state of revision and readjustment, and the hung ceiling does not answer that problem at the local working location. In Fig. 11-2, one solution is shown as developed by the H. H. Robertson Company. H. H. Robertson, developers of Q-deck, have now designed an integrated floor conduit with a high degree of flexibility. In one direction, the Q-deck spaces are utilized for power and communications feed, and right-angled ducts are added in the floor. Robertson indicates that the Q-floor taproute system can normally be amortized in about 5 years in savings from wiring new work stations alone. Naturally, the life-cycle savings are a function of the number of changes required. If the office location remains virtually unchanged over the lifetime of the building, which is

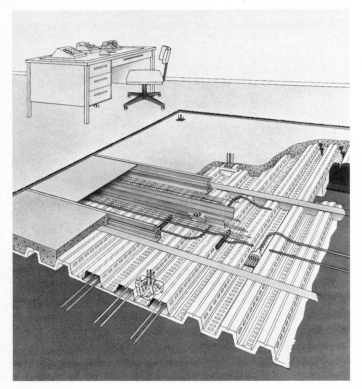

FIGURE 11-2
The Q-floor taproute system (subfloor view). (H.H. Robertson Company.)

unlikely, there would be minimum savings and the initial investment would be at a premium. Choice of a flexible floor system at this stage of the design is an important value-analysis decision.

In performing an analysis of this type of system, Hudson Associates of Washington, D.C., value specialists, have considered alternative approaches. Where the requirements for office station flexibility are substantial, the cost of a complete system could easily be 5 to 10 percent of total initial cost. Recognizing a limitation in the standard systems which limit the number of outlets available at right angles to the main flow, Hudson Associates developed an experimental charged floor system made up of steel plates which are pierced with a special device at any position. The plates are actually a sophisticated floor sandwich which is tapped to provide both communications and power at any location. A number of factors such as the potentially high induction load at loss of total power has been investigated and found to be well within acceptable boundaries. Intuitive knowledge would have indicated that the plates should have been made of a conductor such as aluminum or copper. However, the high cost and softer wearing qualities of those materials coupled with a large square-foot area available resulted in the selection of steel after a value-engineering study. The conductance of a material is a function of its area times the conductance factor, and the large sheets of steel provided ample conductive capacity. This system is still in the development stage, but field tests have been conducted by Hudson Associates and the results are promising.

Hudson has proposed another more immediate alternative to the logistical distribution problem by suggesting that false or raised flooring similar to that used in computer centers be considered for the distribution space. In a sense, this is an upside down hung ceiling. The cost is greater because the support now must be substantial enough to support the live load applied as well as the weight of the raised floor itself. Also, the floor surface must be equivalent in wearing capacity to standard resilient flooring or carpeting. Proven products are available today to meet these requirements at a cost of about $5 to $7.50/sq ft as the result of years of development in the installation of computer floors.

The raised-floor approach is relatively expensive but provides a multiple-function advantage. Utilities can run in the space serving the floor above and the floor below. In addition, maximum flexibility is available for power and communications distribution. Also, the secondary structural problem of providing a false floor is relatively straightforward, as has been proven by computer floor installations.

Studies on a value basis as described in *Architectural Record*

(Ref. 2) have led to even more direct solutions, based upon plug-in raceway systems with power/communication poles to carry the local feed from the hung ceiling to the working station.

One example of such a system is produced by Microflector Company, Inc. It is a prewired raceway, comes in 12-ft lengths and is manufactured to replace one of the main tees of a standard lay-in acoustical ceiling. The raceway has integral receptacles to receive plug-in heads, and a splice box connects secondary feeders from panel boards to the raceways. A system of this type was developed for Sears Roebuck by the Sears corporate architect, Wesley King, Skidmore Owings & Merrill. This was part of the development of a prototype store program in which a modular approach was developed consisting of a precast concrete shell enclosing 30×40 ft steel-framed bays, roof-mounted HVAC systems, prewired electrical distribution (ceiling), and extensive use of downlighting. While this system clearly provides a good cost-value ratio, the individual application must be evaluated, and any field problem which the combination of crafts (electrical and hung ceiling) and/or building-code restrictions might add must be considered. Note the additional flexibility which the plug-in modules provide for relocation of ceiling light fixtures.

A similar system, manufactured by Hauserman, Inc., utilizes a flexible metal conduit connected electrically to a junction box in the ceiling. While less flexible, this approach would have less problem in meeting building codes in most urban areas. Installation of the system is essentially equivalent to standard installation, with complete flexibility for relocation. As described in Ref. 2, the columns contain up to three power lines plus other low-voltage systems (telephone, clock signal, speakers, etc.). Screen-mounted power columns have one incoming power line plus additional low-voltage systems. The degree of modular flexibility is determined to a degree by the spacing of the junction boxes in the ceiling plenum.

Owners, particularly of office buildings, schools, and hospitals, have recognized the life-cycle cost of renovations over the life of the building use. The utility distribution system is one problem area and is readily solved by use of systems described in this section. Another major cost is the moving of partitions. Recognition of this cost has made the use of demountable partitions desirable in many situations where major rearrangements are anticipated at a regular frequency. In schools, there has been a strong tendency toward instant rearrangement through the use of folding partitions, curtain systems, and even open areas which are acoustically designed to be noninterfering. Office landscaping has become increasingly utilized in this context.

Flexibility is expensive. A definite price must be paid for it, and careful value analysis should be made of the true potential for relocation before this investment is made. Flexibility requires long-span structural members and non-load-bearing partition walls, which if not utilized dramatically increase the cost without recovery. Where the relocation might occur once, twice, or even three times during the life of a structure, masonry walls can be a very effective semimovable partition. Many excellent proprietary demountable wall partitions are available. At the preliminary design level, careful analysis of the overall function of the space is necessary. For instance, in one academic building, the designer had decided upon acoustical treatment of the floors as well as the ceiling in room areas. Walls were to be demountable, but since this meant total flexibility in the future, corridors were carpeted. It was decided that partitions should mount on top of the carpet so that future relocations would not cause strips in the flooring which would be difficult to maintain. These decisions led to certain cost and time consequences. As far as cost, the carpeting in any part of the floor could potentially become corridor carpeting, and therefore a class A fire-rated carpet had to be utilized which cost more in terms of time and money to procure. Secondly, since the partitions were to mount on top of the carpet, the building had to progress to a greater stage of completion, including temporary heat, before the carpet could be laid. This delayed certain processes which might have gone ahead under other circumstances. This does not suggest that any of the criteria selected were improper, but rather that they had definite value consequences and that part of the preliminary design included the value evaluation of these decisions.

STRUCTURE

During the schematic design phase, the basic structural system has been selected. In the preliminary design, a great deal more definition is added to the structural system. In fact, in phased construction, the structural system has often been completed in its design stages by the end of the preliminary design. The modular analysis of the spatial solution may have direct impact upon the structural. For instance, the utility distribution system could lead to a decision to provide totally flexible interstitial spaces, although this decision would usually have been made during the schematic phase. During the early stages of the preliminary design, major revisions may occur, and these in turn could recycle the structural thinking which was approved for the schematic phase.

Hudson Associates, Washington, D.C., addressed the problem of installing a waffle panel slab indicated both by the floor loading required and the finishing effect desired by the architect for the ceiling.

In addition, the owner and designer wanted a maximum number of floors within a limited total height restriction, which, according to the Washington D.C. building code, had to be less than that of the Capitol. The waffle was to be closed off by a troffer-type light fixture. This would, under normal circumstances, have made the waffle space unavailable for any other use. Hudson Associates developed a means of piercing the waffle web by the use of a special plastic insert placed between the waffle forms.

The result was a lighter weight concrete with no loss of strength because the plastic inserts were in the web. Appearance was essentially the same as a complete waffle, and the system had four-way flexibility for the inexpensive conducting of small ductwork or lighting conduits and any other utility required.

Value analysis for reinforced-concrete structures should encourage realistic savings. For instance, intricate detailing of reinforcing bar might produce a paper saving in reinforcing steel. However, if the spacing and sizing of the bars undergoes a substantial variety of changes, the designer will have to include a greater safety factor, recognizing that the field-installation personnel will make more errors. Field installation will be slower and more expensive. The overall result could well be a loss in time and money. The reinforcing steel solution should recognize the difficulties of placement of rebar in the field. It is quite possible that an additional investment in steel material will result in a field savings and an overall savings.

In designing structural steel, value analysis can produce savings through the imposition of good practice guidelines, such as:

1. Use of standard structural sizes. The use of unusual sizes will increase mill waiting time because special runs are required.

2. Limited use of special shapes. Again, special runs are required, and the special shapes may not be available in warehouse inventory. If they are, a premium will be required.

3. The variety of sizes and shapes should be kept to a minimum, since an entire mill order may be sidelined waiting for the completion of one or two special shapes or sizes.

4. Development of a member list, with shipping lengths grouped into one or a few reasonable sizes. This is accomplished by adding smaller members together in preparing the size list.

5. Use of standard column sizes. On a mid-rise or high-rise structure, this can be accomplished by using higher grade strengths in the lower floors. The effect is the introduction of standard modules of framing. This makes shop drawings easier to prepare and understand. This, in turn, reduces shop fabrication time and field erection

time. This is another case where purchase of a slightly greater amount of material can produce field savings.

Bethlehem Steel confirms that a designer achieved this type of saving in 14 bulk mail distribution centers. Giffels Associates, Detroit, prepared the plans and specified standardized building columns on 14 projects to facilitate modular design. The approach permitted uniform detailing of mechanization equipment support steel. Even though the heights and spans on all 14 projects were kept the same, the balance of the steel framing was designed for actual weather and equipment loads at the individual sites. The modular approach, in turn, simplified the sizing of building machinery.

BUILDING EXTERIOR

The enclosure of the building exterior is a question of functionality as well as aesthetics. Natural light has a substantial psychological impact. Until recently, the majority of working space in office buildings, schools, and hospitals had to have exterior light. When building ventilation was achieved by the use of double-hung windows, natural light was available at no extra cost. Now that almost all major buildings are operated on a closed system for heating, air conditioning, and ventilation, the provision of natural light is expensive in terms of the first cost and life-cycle cost.

The International Masonry Institute points out that in hot weather in Washington, D.C., the heat gain through a square foot of insulated brick and concrete block wall will be 2.2 Btu/hr, while a double-plate glass wall in the same location will gain 173 Btus, or almost eightyfold more. In terms of cooling capacity, a 2-ton capacity could handle a heat gain through the masonry wall, but 143 tons would be required for the glass wall. These figures were developed on the basis of a heat transmission U factor of 0.12 for the masonry wall and 0.55 Btu/(hr) (sq ft) (°F) temperature differential.

The masonry wall described has a 38 percent lower initial cost than the comparable double-plate glass wall. In terms of life-cycle costing, for a 10-story building the masonry wall would cost $0.23/sq ft over its life versus $7.60 for the double-plate glass wall. Over that same life, the heating cost per square foot for masonry would be about $0.30 versus $1.38 for the glass. As fuel costs mount, these comparisons will become even more divergent and dramatic. As indicated in Table 11-1, the comparison is certainly a fair one, since the U value for single-pane window door and store front glass is 1.09.

In the *Engineering News-Record* (Ref. 3), the Illinois Capital Develop-

Table 11-1
Heat-Transmission Factors, U (U = Btu/hr/sq ft/degree)

Material	Typical U factors
Exterior walls:	
Glass, single pane	1.09
Glass, double-vacuum space	0.64
Glass, double-vacuum, metal coated	0.50
Stone, limestone 8 in. plain	0.71
Concrete, 10 in. monolithic	0.62
Brick, 8 in. plain	0.50
Hollow tile, stucco	0.40
Brick, 12 in. plain	0.36
Brick 8 in., plaster on lath, furred	0.30
Frame structure, siding sheathing, full insulation, dry wall (approx.)	0.10
Insulation:	
4-in. cavity wall	0.09
3-in. polystyrene foam	0.07

ment Board, which handles about $500 million in construction annually, discussed certain of its new energy requirements. Illinois elected not to use a goal of total utilization per gross square foot by citing 55,000 Btu/(gross sq ft) (yr) as maximum consumption. Illinois considers that this would be too difficult to monitor and control. Instead, Illinois is calling for a U factor of 0.23 for gross exterior walls. This is readily achievable with masonry walls. A key factor is the percentage of fenestration. Ref. 3 notes that some all-glass buildings have a gross area U factor of 0.75 and even higher.

The Illinois U-factor levels for new structures call for 0.10 for gross roof area, 0.10 for opaque wall sections, and 0.06 for opaque roof areas. The FHA requires that masonry walls for multiunit housing will have a U factor no higher than 0.17. The city of New York has indicated that its building designs shall have 25 percent or less window area to encourage energy conservation.

The question of exterior walls is key to life-cycle energy costs. Value analysis in these areas may well lead to new methods of achieving desired results. The primary function of glass windows is not really the admission of natural light as much as the preclusion of the psychological effects of a windowless building. In answering the exterior problems created by windows, many designers have gone to the total window wall.

For a low-rise high-quality office building to house a major corporate headquarters, the architect selected full-height window walls. In deference to the energy considerations, tinted, reflective, insulated glass was specified. The typical window-wall section is shown in Fig. 11-3a. Note that the desired exterior effect of full-height glass produced an awkward interior detail at the hung ceiling intersection with the wall. The solution was an expensive offset to the structural spandrel.

The owner-developer management team reviewed the detail and revised it, as shown in Fig. 11-3b.

By adding a mullion at the same height as the hung ceiling, the framing of the wall-ceiling intersection was simplified, producing a modest cost saving.

More importantly, the additional mullion allowed the reduction of one size in glass thickness at an initial savings of 20 percent in square foot cost. Further, the owner team replaced the upper section of the window wall with an insulating panel rather than glass. This was a substantial reduction in first cost.

The most noteworthy feature of this savings of about 33 percent in initial cost (almost $300,000) was the significant life-cycle impact.

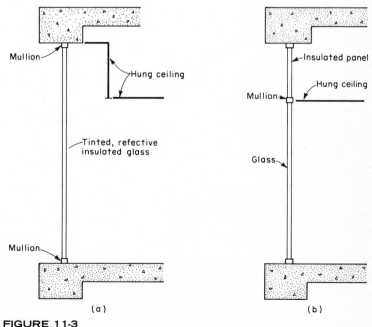

FIGURE 11-3
(a) Typical window-wall section; (b) revised typical window-wall section.

First, the exposed heat-transmission area was changed from one maximum zone to two zones.

The lower zone (67 percent of original area) continued to transmit heat (at a faster rate because of the reduced window-wall thickness). However, this increase in heat loss was more than offset by the change in the upper zone (33 percent of original area). The combination of lower heat gradient in the hung ceiling and insulated panel dramatically reduced the overall heat loss of the system.

Perhaps value analysis will lead to the utilization of already-developed systems such as the National Gypsum Company's metal-edge core wall-glass window wall. In a telephone equipment building which did not require windows, Chicago architects Holabird and Root designed a wall sandwich made up of inexpensive 2-in. metaledge core wall backed by 1-in. urethane foam and $5/8$-in. gypsum wallboard. The overall package had a low U factor and also had a low initial cost. To protect the wall system from the elements, the exterior closure was completed with a single pane of exterior glass. The sandwich had the advantage of low heat transmission, while providing from the outside an attractive window-wall appearance. For a building requiring the admission of natural light, windows could be created easily in the interior portion of the window sandwich. In terms of heat transfer, a tremendous reduction could be made in percentage of fenestration, while from the exterior the desired aesthetic effect could be achieved at much lower cost. Costs involved in total window walls are not limited to the cost of the window wall itself but also include additional requirements for draperies, curtains, and shades to permit the exclusion of sunlight, taking into consideration changes in season and specific location of the building.

In low-rise buildings, heat transmission through the roof is a substantial energy cost factor. Owens-Corning points out that specifications of $2^{1}/4$-in. fiber glass roof insulation instead of a thinner $15/16$ in. can save up to 50 percent of the energy loss through the roof. A secondary advantage is reduction in the amount of heating and cooling equipment. Owens-Corning estimates that for every dollar spent on insulation, reduction of up to $2 of original equipment costs can be achieved. For suburban office buildings in the northern climates, this type of increase in insulation could save as much as $27,000 in equipment costs for every 60,000 sq ft of roof.

The Grace Construction Products Division of W. R. Grace Co. estimates that the use of masonry fill insulation in cores or cavities of masonry walls can reduce heat loss through the walls by 50 percent. Table 11-2 shows a cost comparison on savings from insulating walls in a two-story office building over a 10-year period. The return on investment is dramatic.

Table 11-2
Insulation Cost Savings

	Chicago	Atlanta	Mpls.	Phila.	Denver
Combined heating/cooling savings*	$6400	$3500	$8150	$6450	$5400
Installed cost of insulation	1700	1700	1700	1700	1700
Average annual return on insulation investment	38%	21%	48%	38%	32%

* 10-year savings from insulating walls; 8-in. lightweight block; 2-story office building, net exterior wall area 10,000 sq ft.
SOURCE: W. R. Grace Co. from *Architectural Record*, May 1973.

The high energy cost of materials such as glass will inevitably lead to a greater utilization of masonry because of its lower heat-transmission value. This change in emphasis could well lead to more load-bearing concrete masonry. Presently, masonry is used principally as an exterior in the structural system, and not in a load-bearing capacity. However, load-bearing capacity can be developed with little or no additional cost so that for the range of $2 to $6/sq ft, structure *and* the majority of the exterior wall system can be purchased. Load-bearing masonry walls have been used in mid-rise and even high-rise structures, although the practice is much more prevalent in Europe and the United Kingdom than in the United States.

HVAC

During the schematic phase, the HVAC systems can be defined only to a limited extent. The basic load factors can be developed. For instance, in Dallas, the first International Building estimated that its loads were as follows:

Heating	7%
Equipment	14%
Cooling	29%
Lighting	50%

The balance between loads leads to certain basic decisions in the selection of the HVAC systems. In northern areas, the systems can be designed to utilize lighting load as a heating source. In the Dallas building, heating is the least of the problems, with removal of heat being a much more substantial consideration.

The design of the heating, ventilating, and air conditioning systems requires evaluation of all factors in the building, in particular the criteria for HVAC performance, the cost of energy, and the configuration and materials which make up the building exterior. An overall cost model is developed in relationship to a total heating-cooling balance equation for all HVAC systems. The mechanical and electrical engineering firms specializing in HVAC have developed a number of computerized systems for setting up and evaluating systems. This is important to value analysis, since it provides a definition of the viable alternatives. Energy and equipment companies also provide this type of capability. The Southern California Gas Company makes available a computer system called E-Cube which performs a computerized energy analysis. E-Cube is a three-part program that first calculates the energy requirements, considering hour-by-hour building use and occupancy profiles. This requirement profile considers heat losses and gains, weather variations, solar exposures, building orientation, control sequences, structural conditions, and the electrical and thermal load peaks and valleys projected throughout the year.

The second part of E-Cube compares four different HVAC systems from all electric to total energy, evaluating the performance of each system by simulating its operation in the context of the detailed energy requirement program. The system identifies each alternative system's thermal efficiency, potential for recovery of waste heat, and sizing of units, and it provides a printout which summarizes the monthly and annual gas, auxiliary fuel, and electrical demand and consumption.

The third part of E-Cube compares the alternative systems on a cost basis, including initial investment, annual depreciation, operating maintenance and replacement cost, methods of financing, rate of return on investment, and cash flow before and after taxes.

The Trane Air Conditioning Company has developed a computerized analysis system called TRACE (Trane Air Conditioning Economics). This system was used to analyze a proposed building with the following parameters:

20-story building: 300,000 sq ft

Cash flow first year: $125,675

Gas and electrical energy

Absorption water chiller

Double-duct air conditioning system

50 percent insulated clear glass

Lighting intensity: 4 watts/sq ft

East/West primary exposures

3-ft insulation

18 cu ft/min per person ventilation

Trane measured the advantages of recommended changes by a net cash-flow (annual) figure. This figure is the combination of any additional increase in cost as balanced by cash savings in utility costs at current rates. The recommendations and net cash-flow savings were:

1. Change type of energy use for heating	$ 5,135	4.1%
2. Change from absorption water chiller to centrifugal water chiller	2,343	1.9%
3. Change from double-duct to variable-air-volume system	11,449	9.1%
4. Change from clear glass to reflective glass	15,221	12.1%
5. Change amount of glass from 50 to 20 percent	16,932	13.5%
6. Decrease lighting intensity 1 watt/sq ft	9,089	7.2%
7. Rotate building by 90 degrees to change solar exposure	4,805	3.8%
8. Increase insulation by 3 in.	3,395	2.7%
9. Decrease intake of outside air from 18 cu ft/min per person to code requirement of 7.5 cu ft/min per person	2,061	1.6%
	$45,874	36.4%

The net cash-flow improvement was principally a result of an annual utility-cost savings of $63,354. Obviously, energy savings were directly related to substantial improvement in carrying costs. In any individual value-analysis study, the cost of interest, area of the country, energy costs, and other factors would vary substantially.

An important factor in the implementation of savings for systems such as variable air volume is the ability to actually cut back on systems not required. At the University of Cincinnati, a Honeywell Delta 2000 central control system computer was used to cut running time by 40 percent. The computer automatically starts and stops heating and cooling systems, depending on space occupancy. The director of the physical plant estimated that the savings in energy costs over 2 years were $242,820.

Other cost savings may be achieved by placing the HVAC unit directly in the area where it is needed. This principle has long been used in industrial parks with rooftop units. While central supply costs less per ton of air conditioning or Btu of heat generated, the distribution system often more than offsets this efficiency. Again, the value analysis must be based upon the actual situation and the parameters. In the Trane analysis above, certain of the savings were realized by common sense revisions in the parameters of performance. High-intensity light levels are not necessarily helpful or directly related to higher efficiency beyond a certain optimum point. Fresh air turnover should be carefully considered, since it is expensive. Reutilization of preconditioned air with recycling of contaminants may be much more cost effective. Energy conservation techniques are being furthered through the use of heat wheels which can retain heat or reject it, depending on the appropriate cycle and season.

Other special equipment which is now available includes automated energy-recovery systems which can extract heat from waste materials, thereby solving two problems at one time. However, the utilization of recovery systems must be integrated into the total HVAC solution.

ELECTRICAL DESIGN

In the preliminary design phase, the electrical work still depends to a great degree upon the definition of the HVAC design and the lighting intensity and layout determined by the spatial solution. The lighting layout must follow the spatial layout and is relatively easy to design. The selection of the type of fixture and the intensity provided can be a high-value decision. The recent studies have indicated that there is a tendency to over-design in terms of the wattage per square foot. In a recent GSA design for the Saginaw Federal Office Building which was mandated as an energy saving approach, the designers by careful placement of the light fixtures were able to achieve a 65 foot-candle lighting level with approximately 2 watts/sq ft rather than the 4 watts/sq ft, which would have been considered standard.

During this phase of the design, the electrical engineer can go into considerable depth in regard to the electrical service and the type of source. At this stage of the design, the electrical engineer must select a standard power source, total electrical energy source, or the composite approach often termed selectric. This decision will determine the primary service. The secondary transformation and distribution requirements will be determined by the overall loads and their locations.

In *Architectural Record* (Ref. 4), Joseph F. McPartland, editor of *Elec-*

trical Construction and Maintenance, described the considerations which the electrical engineer has to keep in mind during this particular phase of design:

> Distribution of electrical power in commercial and institutional buildings involves a sophisticated and sometimes complex combination of electrical technology, commercial and economic considerations, and diligent attention to safety for persons and property. In system design, the electrical engineer must observe the general principles that will help insure the system meets the owner's utilization needs. And, equally, he must observe technical details of layout, design and installation to meet modern codes and standards to provide safe, effective operation of all segments of the overall electrical system. But electrical systems cannot be designed by code alone, because the designer must consider factors not fully covered by code such as voltage drop, power factor, detailed analysis of watts-per-square-foot loads, demand factors, and provision of substantial spare capacity.

Local and major codes, such as the National Electrical Code, give very specific information in regard to required conduit size and wire size for specific loads. These codes are now being comprehensively enforced and provide basic guidelines for any value analysis.

Safety considerations as well as insurance requirements insist that codes be complied with totally and that installations be inspected by certified inspectors for compliance with the National Electrical Code.

Reliability is a vital factor in the operation of modern buildings, and the basic design must include the back-up sources—either commercial or emergency power. The selectric approach, which has a combination of each, automatically provides emergency power without a specific emergency generation system. This is a good value, since the equipment does not merely stand idle but has a functional utilization.

In combination with source for emergency loads, new electrical design considers the need to shed load during high demand periods. The facility must sense high voltage drops in these periods of brownouts and load shedding by the electrical utilities at the primary source.

The electrical system should provide the flexibility for growth. Inevitably, the load grows or is required at different portions of the facility. Judicious provision of spare capacity to facilitate this growth can be an important value purchase, and similarly, inappropriate inclusion of back-up or spare capacity can be a poor value selection.

In an office park structure, the electrical designer added certain sophisticated relay protection to the main switchgear. From the viewpoint of the functions of the building, including a computer center, the additional features appeared a good investment. The owner's value-analysis team deleted the features when they determined that a reliable alternative source of primary power was not available. The office park was served by a single aerial high voltage line. The electrical-utility planning division advised that alternative main service feeders were not in the capital plans for the next 10 years. The initial cost savings were $40,000, with better delivery time.

In Ref. 4, McPartland discusses new techniques in design which must be considered:

Equipment Standardization—maximum standardization of equipment type and rating can effect significant savings because standard equipment costs less than special equipment and replacement parts are easy to get.

Overload Protection—because of the continual increase in building electrical loads, larger and larger service entrances are being fed from utility supply systems that have greatly increased levels of available short-circuit currents. Potential destructiveness (fire, explosions) must be held in check by over current protection.

Ground-Fault Protection (GFP)—faults are aberrant conditions in which currents can take wrong paths and create hazardous conditions such as fire or potentially fatal shock; a fault between a conductor and ground is commonly called a short circuit. Arcing ground faults occur more frequently nowadays because of the use of higher-voltage distribution in buildings. Particularly at 480/277 V three-phase distribution, an arc can develop between one phase and ground, possibly drawing insufficient current to trip a current-protective device, but creating a sufficient arc to start a fire.

Ground faults at lower voltages may not develop an arc, but can energize a presumed grounded surface, such as an appliance. Enough current can flow to be fatal to a person without a conventional protective device being tripped.

Both of the above conditions have led to the development of ground-fault interrupters that sense small leakage currents.

The NE Code requires ground-fault protection on grounded-wye, 480/277 V electrical services where the disconnect is rated at 1,000 amperes or more. Also GFP is required for outdoor residential circuits, at construction sites, and under some conditions, for swimming pools.

Recognition that the basic use of watts per square foot or standard foot candles really does not guarantee the delivery of proper lighting levels for specific tasks has led to the concept of equivalent sphere illumination (ESI). This complex approach coordinates the location of luminairs, geometry of the luminair itself, room geometry, distribution of the light from the luminair, room reflectances, and the impact of horizontally polarized light. When these factors are combined in complex formulas, ESI foot candles can be measured. However, because of the complexity of formulation, computerized evaluation must be made.

The A/E firm of Smith, Hinchman and Grylls developed a computer program for this purpose and offered the algorithms for the program at the Illuminating Engineering Society National Conference in 1974. An example of the complexity is seen when the SH & G criteria for design are prescribed. SH & G investigates the lighting values for four viewing directions at approximately 400 points into space. The output is shown in the form of a contour plot which maps the equivalue lines.

This additional investment in electrical design can result in a substantially lower first cost and life-cycle cost while delivering optimum lighting levels. The savings projected in one project by SH & G was a reduction in average lighting wattage from 4 watts/sq ft to 2 watts/sq ft, with a subsequent reduction in air conditioning tonnage from 1,600 to 1,350 tons.

REFERENCES

1. *Engineering News-Record,* Apr. 24, 1974, p. 20.

2. *Architectural Record,* Apr. 1974, pp. 155–156.

3. *Engineering News-Record,* May 23, 1974, p. 13.

4. McPartland, Joseph F.: *Architectural Record,* Aug. 1974, pp. 99–100.

CASE STUDY K
PRELIMINARY DESIGN (SELECTIVE ENERGY STUDY—FLACK AND KURTZ)

This case study is extracted from a complete value analysis performed by Flack and Kurtz, Consulting Engineers, New York, after the schematic design phase for a 500,000-sq ft building expansion for a community college.

The firm of Flack and Kurtz, very concerned with the delivery of optimum value, is a participant in the design of a prototypical office building which will produce an optimum balance between initial costs and operating costs. The building concept has been designated En-Con.

The community college expansion would approximately double the size of the existing school complex, which already had a power plant and distribution system. The analysis had to recognize the value and operating potential of the existing plant, and develop the optimum approach for providing additional heating and cooling equipment and meeting lighting and power requirements.

The three basic approaches to the development of the additional power requirements were:

1. Basic Scheme This would basically be an expansion of the existing high-temperature-water (HTW) heating system by provision of additional boiler capacity. Electrical power would be purchased from Con Edison, and additional cooling capacity would be provided by absorption machines.

2. Selective Energy Scheme This approach would also provide HTW heating and absorption machine cooling. Power for the existing 1,000 kW of load would continue to be supplied by Con Edison. The additional requirement of 2,500 kW would be generated on site. (This approach, previously described, was implemented by the New York Dormitory Authority and the City University of New York at the Richmond College Campus.)

3. Total Energy This approach would generate all electric power requirements on-site.

Evaluation of the basic scheme indicated that addition of one 18000 MBH (thousands of Btu/hr) generator in addition to the three existing 14000 MBH HTW generators would provide the required capacity, with any two of the existing boilers on line and one for standby.

Additional electrical capacity would be added to the existing Con Ed system by connecting supplementary feeders to new substations around the site. In addition, 600 kW of emergency generator capacity would be required.

The selective energy scheme would add three gas turbine generators, two to carry the base load and one for standby to meet the new load of 2,500 kW. Waste heat from the turbines would generate high temperature water to meet a large part of the heating and cooling loads. Standby in the balance of the requirements would be met through the existing boilers.

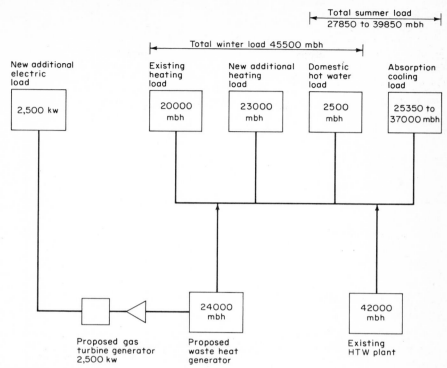

FIGURE K-1
Selective energy scheme flow chart.

Emergency generator capacity would not be required under this scheme. The load-balance diagram is shown in Fig. K-1.

The total energy approach would be essentially the same as the selective, except that the load would now be 3,500 kW, requiring four gas turbine generators with three carrying base load and one for standby. The entire campus would be served by this approach, and Con Ed service would be disconnected. Emergency generator capacity would not be required. Table K-1 shows the initial installation costs for each scheme.

Annual loads were evaluated on a degree-day basis and costed using current projected figures for oil, gas, and electric power. There was no way to project future increases, and so the assumption was made that increases would be proportional.

Table K-2 compares the annual operating costs on three bases. First is the cost developed at an earlier point, using existing rates and oil as a basic fuel. Column 2 shows the current scheme, again using oil. Column 3 is a dramatic change and is based upon favorable gas rates. Recognizing that the parity between gas and oil will continue to fluctuate in the future, the

Table K-1
Energy Supply Alternatives First Cost Summary, $
(Oil and Gas)

Scheme	Basic	Selective energy	Total energy
Boiler and auxiliaries	68,000	—	—
Gas turbines with waste heat boilers	—	770,000	1,064,000
Building space	18,000	160,000	200,000
Additional electrical construction costs	—	120,000	45,000
Installation	30,000	190,000	216,000
Emergency generators	240,000		
Conversion of existing burners	75,000	75,000	75,000
Total	431,000	1,315,000	1,600,000
Design contingency 10%	43,100	131,500	160,000
Total	474,100	1,446,500	1,760,000

engineers recommended that there be a method for converting from gas to oil wherever possible.

Basing their recommendation upon these figures, the engineers indicated that either selective energy or total energy would be a favorable economic choice. It was recommended to adopt a selective energy scheme which would maintain the existing Con Ed service connections while providing the advantage of separate sources of energy and, therefore, greater flexibility.

A key to the feasibility of the selective and total energy approaches on this campus is the use of absorption cooling equipment, which provides a market for the waste heat during the summer months. The economic balance would be completely destroyed if standard electrically driven chillers were utilized. The higher cost of the absorption equipment has been factored into the overall value-analysis model.

Table K-2
Comparison of Annual Operating Costs*

	Old rates (oil)	New rates (oil)	New rates (gas)
Basic scheme	$996,200	$1,301,295	$1,215,425
Selective energy	$997,920	$1,150,906	$987,631
Total energy	$991,020	$1,058,670	$855,820

* Includes energy, maintenance, operating differential, and amortization of capital cost.

DETAILED DESIGN PHASE

The detailed design phase takes the design concept which has now been defined in the design development and adds the level of detail required for the contractors to construct the facility. Major components of this portion of the design include:

1. Layout of partitions and interior fixtures
2. Engineering design of structural elements at the detail level
3. Architectural design of construction detail
4. Layout and design of mechanical systems
5. Location of fixtures and utility outlets
6. Architectural design of interior and exterior features
7. Chart of materials and finishes
8. Draft of specifications
9. Coordination, review, and check of drawings
10. Final working drawings and specifications
11. Preparation of proposed contract, including general and special conditions to be followed

In addition to the specific design and contract components, this stage of design often includes a detailed cost estimate. In terms of value analysis and control, this detailed cost estimate is much too late. The best opportunities for major evaluation and change of direction were, of course, in the earlier design stages.

The final phase of design is the preparation of the contract documents. The work involved is a function of the type of contract which is contemplated. If the contract is to be negotiated and/or the contractor is to receive all his costs plus a profit, the contractual documents can be relatively brief. In fact, in this type of contractual relationship, a performance specification will work quite well. This specification describes the level of performance anticipated by the owner, and in a negotiated environment the owner can always call for a better quality, or lesser, as best suits the purpose.

The majority of building contracts are of the lump-sum type, based upon a competitive bid. In order for the bids to be truly comparable, the bidding documents must include a set of drawings and specifications which are very definitive. Preparation of this category of detail and the contract documents requires a design time at least equal to the length of the schematic and preliminary design phases combined.

The detailed design phase is very structured in its scope. The preliminary design, if properly implemented, defined the design. In effect, the skeleton has been established, and the detailed phase is filling out the detail. Most of the major value analysis should have been accomplished in the prior two phases. However, while the breadth and scope of value analysis is thus limited, it is still practical to carry out comprehensive and effective value analysis during the detailed design phase. The value analysis subjects should be compatible with the breadth of the detailed design scope. While it is possible for major changes in thrust to occur during the detailed design phase, these major changes would inevitably be costly. As a rule of thumb, a delay cost of 1 percent per month should be contemplated. This represents an out-of-pocket cost for escalation and inflation and does not compensate the owner for the loss in the utilization of the facility—which is probably also at least 1 percent per month. Obviously, major conceptual analysis changes should be recommended during the schematic phase, and major design elements should be recognized in the preliminary phase.

There are occasions where a major concept can be changed without impacting prior decisions. For instance, in a sewage storm drain system, a new 3,622-ft relief storm drain replacement was specified. The contractor included an alternative lump-sum price of $22,000 to line in place an existing portion of the present corrugated metal pipe system. The township accepted the alternative, and the saving was estimated at $23,000 for lining of approximately 200 ft of 60-in. pipe and a similar amount of 30-in. pipe.

There are three major areas of value-saving potential in the detailed design phase:

1. Revision of materials to produce either more value or less cost

2. Specification of installation methods to incorporate labor-saving methods

3. Review of contract requirements to delete redundant or excessive requirements

MATERIALS

Structure

In reinforced concrete portions, it is often possible to specify a lesser-strength concrete for certain areas. Overspecification of concrete strength requirements is often the result of habit on the part of the individual designer. Similarly, concrete finishes are often specified at more than the actual requirement. For instance, a trowelled finish may be called for in a portion of the foundations which are to be backfilled. In other cases, a trowelled finish may be called for where safety would prefer the less-expensive wood float finish.

In highways, a substantial saving has been realized by the use of continuous reinforcement. This reduces or deletes the requirement for expansion or contraction joints, with a resulting improvement in value and performance at a lower initial cost. The galvanizing of reinforcing bar for highways can result in a longer useful life at a slightly increased initial cost.

While the major structural decisions will have been made in the earlier design phases, the detailed questions of finishes and connections can be treated and analyzed in the detailed design. A very key value judgment has been made in the industry favoring the use of high-strength bolted connections for field erection. This has displaced riveting to a great extent, with a better-quality joint at lower cost. The shop fabrication phase has not followed suit; the principal reason is conservative traditional thinking.

Building Exterior

The basic method of providing the building exterior skin will have been selected in the preliminary design phase. This leaves several options still open in the detailed design phase. In a masonry wall, for instance, a larger block might be selected. Value analysis might disclose that the use of a composite block with a finished exterior could be less expensive than a finished stone, block, or brick with a masonry back-up requiring two separate operations.

Exterior wall solutions have a concomitant interior wall surface to contend with. Accordingly, an inexpensive approach could dictate a more expensive interior. Some curtain wall panels are constructed with a complete inner and outer surface which could save money by the reduced number of installation operations.

In selection of window units, size is a factor in determining whether units can be shipped preglazed. Preglazed units can materially reduce field-installation time and can provide better overall quality control.

Roofing

Basic roofing materials have not changed over a number of years. However, some new ideas in the methods for configuration for installation are evolving. Hudson Associates is studying the placement of wildlife scenes on building roofs, including duck ponds in sylvan settings. The idea evolved from observation that in Europe it is quite usual to have sod roofs or sod on top of roofs.

Naturally, for any such installation the basic roof placement must be sound. The reason for installation of these attractive scenes is for life-cycle improvement of the roof life. Roofing deteriorates principally because of freeze-thaw cycles, exposure to heat, and other weather extremes. The Hudson idea is to insulate the roof from the elements while providing an aesthetic improvement in rooftop appearance.

Figure 12-1 shows a less picturesque solution to the same problem: the IRMA (insulated roof membrane assembly). This concept reverses the usual sequence of insulation of a roof. The three-ply membrane is placed next to the roof deck and is topped by Styrofoam RM insulation. The insulation on top of the roof serves two purposes: (1) it con-

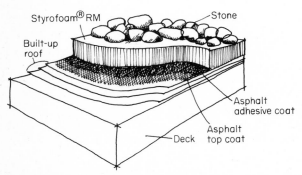

FIGURE 12-1
The insulated roof membrane assembly (IRMA) concept.
(Courtesy of AMSPEC, Inc.)

tinues to insulate the interior of the building from the elements in terms of temperature, and (2) it mechanically protects the roof from damage, while also insulating it thermally. As a result, the roofing is in a protected environment and survives to function for many additional years.

Piping Systems

There has been a longstanding conservative tendency to overspecify the piping systems. In some cases, this is the result of existing building codes, and there is no substitution permitted. In many cases, however, the specification is the result of the habits which a designer has. Value analysis can ask why a certain level of material is required. For instance, corrugated pipe or spirally welded pipe may be very functional for certain services and considerably less expensive in terms of material and installation than a schedule 40 steel pipe would be.

The scope required for installation of special piping, such as acid-proof piping for classroom laboratories, should be carefully reviewed. Expensive glass piping may be specified where terra cotta pipe could perform the same function at lower cost.

Interior Finishes

Paints specified may have a low initial cost, particularly for material. However, for hard-utilization areas such as academic hallways, an epoxy-based tough finish may produce much lower life-cycle costing. Special finishes specified for equipment may dramatically increase costs and again may be the result of habit in specification. Specification of a grade of carpeting substantially less than the grade required to provide a reasonable life can result in replacement in the first several months at a very high life-cycle cost; the value analysis of details does not always result in lower initial cost or selection of lower-grade materials.

During the construction phase, it is usual to specify furnishings which are major in nature as part of the construction, while the more portable furnishing is part of a follow-on decoration phase. Items purchased under the construction contract tend to be more expensive per unit, such as built-in furniture versus stock furniture, and so the delineation between construction phase and decoration phase should be carefully ascertained. In some cases, the choice is based upon capital versus noncapital costs. In this situation, the value analysis might suggest a redefinition of capital and noncapital funding for the purpose of the specific project.

Careful examination of the finishing schedules may disclose over-lapping. Sometimes, situations will be discovered where two finishes have been specified, as in the case of floors which are first covered with tile and subsequently covered with wall-to-wall carpet.

Lighting

A specification can have unsuspected cost implications. An office building had specifications for two-bulb light fixtures in continuous rows throughout the office space. After value analysis it was recommended that four-lamp fixtures every other 4 ft in the ceiling grid be used instead. The reduction in lighting intensity was only 3 percent, which was considered nominal. However, the elimination of approximately 4,000 fixtures produced an initial cost savings of $25,000. The reduction also permitted movable partitions rather than fixed partitions and left flexibility for altering partitions in the future.

INSTALLATION

Methods of installation have to consider code requirements and unionized trade requirements. For instance, in specifying an integrated ceiling furnished by a well-known manufacturer, one architect did not fully realize the field implications. From a finished point of view, the integrated ceiling lived up to its name. It combined lighting, acoustics, and air distribution. However, the state in which the college building was being erected has a multiple-contract requirement, which means that the general, HVAC, electrical, and plumbing contracts are separate. The integrated ceiling involved three separate contracts and at least five different unionized trades. The result required a working out of the interface between the various trades. Ultimately, the installation cost was 10 to 20 percent more than it would have been in a nonunion environment. While the ceiling was physically integrated, in effect it split the installation trades and thus created unintended interface points and higher coordination costs. The selection of standard items such as knockdown door bucks can reduce procurement time, initial cost, and installation time.

In piping, the selection of mechanical joint pipe for heavy piping systems and systems such as the Tyler Rufwall® system can dramatically reduce field costs. Field unions are becoming more relaxed in regard to their acceptance of this type of advancement, because the unions are becoming more conscious of the need for higher productivity per worker and because the workers themselves relate to ease of installation. From this viewpoint, the craft unions are becoming more interested in the availability of improved hand tools, and in some cases, even in equipment which would previously have required a separate operator, such as hydraulically operated lifts to permit high bay work.

Plastic piping, which has been endorsed by the Department of Housing and Urban Development, is still facing building-code restraint. One of the recommendations of Operation Breakthrough was the development of a single-stack plumbing tree made of vinyl materials. The basic stack would have lower material cost, lower installation cost, and a much longer useful life. This approach has had extensive testing in Europe. It is a preengineered design consisting of a rigid plastic assembly one branch in height, with branches to receive discharge from a complete bathroom and one complete kitchen, including laundry waste. Through the use of special fittings, this branch can be inserted directly into previously installed work at the level below for high rise. Cost reductions of as much as 40 percent are projected over standard plumbing installations. A basic factor in the cost savings is

FIGURE 12-2
Assembly of Tyler Rufwall® connections. (Courtesy of Tyler Piping Co., Inc.)

the ability of the material to be cut with a handsaw. In industrial environments where flameless work is required, a further advantage is the fact that no open flame for soldering or welding is required.

GENERAL CONDITIONS

The standard conditions of the specification often call for accepted practices which may not produce good value. For instance, it is common practice to call for a performance bond. This bond is insurance purchased by the contractor which guarantees the owner completion of the work remaining in any point in the project at the contract price. In many cases, it is mandated by law that the owner shall have a completion or performance bond. For a one-time owner who is building perhaps the only facility that he will ever build, a performance bond is probably a good investment. For the multiple builder who builds many projects, a performance bond can be quite expensive. For each $100 million put in place, the cost of the performance bond is $1 million. For every billion dollars that the federal government places in construction, $10 million is spent on insurance if a performance bond is required. Agencies such as the GSA, Corps of Engineers, and Naval Facilities Command place enough work that self-insurance is worth consideration. The same is true of state governments and major corporations.

Another type of bond which is commonly required is the 10- or 20-year roof bond. This is a guarantee that the roof, once properly installed, is guaranteed over the period specified. The money involved is much less than the overall performance bond, but it is a form of insurance and the owner pays for it. However, implementation of a roofing bond requires a clear demonstration that the roof failed and that it was not made to fail. The owner must show that neither he nor any other party damaged the roof, making it leak. The proof of the matter is often quite intangible, and the execution of a roofing bond is quite difficult. The owner can often save money in the long run by requiring a sound roof but omitting the bond. In case of this omission, the specification may require that the roofing be installed in accordance with a 20-year roof.

Other portions of the general conditions call for a performance of services and/or installation of temporary facilities. Careful planning can often save some of the cost. For instance, if temporary roads are laid in the same location as permanent roads, the temporary road can be constructed at a level of quality equal to the base of the permanent road. For instance, gravel or slag can be placed to provide a construction access road. This saves time and money in free access to the site.

Then, in the final phase, the construction roads can be cleaned off and surfaced with a final wearing surface to provide the permanent road.

Similarly, wherever time permits, contracts should be let to provide permanent utilities which will service the site during construction as well as during the life of the project.

In situations where a number of contractors will be working on the job site, there are many services provided individually. For instance, in multiple-contract situations, the specifications typically say that each contractor will get rid of his own trash and rubbish. Accordingly, in bidding the job, each contractor places an amount of money into the contract for this work. However, on most projects, there is a continuing argument as to who created the trash and who should clean it up. The result is that each contractor attempts to conserve his budget for clean-up, and usually the owner ultimately has to add extra service. A clean job is desirable, and under the Occupational Safety and Health Act, a requirement, and so the owner may be best served by requiring one contract or his construction manager to provide all the clean-up service. The result is an actual savings in cost to the owner and a much better project attitude on the part of all parties.

Temporary heat is an important ingredient for completing the project. At the outset of a project, it is often uncertain how much heat will be required. The general contractor may skimp on this phase to keep his price low, or having successfully won the job, may try to skimp on the implementation of temporary heating to conserve his budget. The designer should consider the possibility of having the owner provide this type of service at actual cost.

This approach of having the owner provide a full service to facilitate the progress of the job is a sound one. It removes an element of gambling from the bid and reduces arguments during the job life. Proper temporary services are an important factor in timely construction. The owner should want these services provided in a timely and appropriate fashion. By picking up the cost during the life of the project, the owner will ensure the willingness of the contractors to receive the service. At the same time, the owner avoids a situation in which he has paid for service in lump-sum form at the start of the project but never actually receives full measure during the life of the project.

CASE STUDY L
DETAILED DESIGN PHASE
(VALUE ANALYSIS)

This case study describes a value analysis through which as much as 50 percent of the fabricating costs for structural steel can be reduced by a change in the detailed design phase. This change involves the connections only and would not recycle changes in the previous design phases.

The information is extracted by permission from technical paper No. 550, *A Realistic Look at Structural Fabricating Costs,* copyrighted by the Metal Fabricating Institute. The material in this paper was prepared by Harry Conn, President, W. A. Whitney Corporation, Rockford, Ill.

The key analytical point presented by this case study is the functional analysis of the structural steel connections. The higher cost and questionable quality of riveted connections resulted in phasing out of that type of connection in almost all building structural steel. The advent of high-strength bolts provided a permanent and convenient field-connection device. This study looked at the shop connection welded to the structural member from a functional viewpoint. The connection of the connectors to the structural members has traditionally been by welding in the shop. The function is "Transmit Force." The field connection of member to member by high-strength bolts demonstrates the permanency and strength of these bolt connections. The study looks at the use of high-strength bolts to connect the clips and angles to the main members to meet the function "Transmit Force."

In presenting this case study, a substantial portion of significant information has been excluded so that the case study can focus on the successful identification of a high-cost item which has been perpetrated by convention:

Design, fabrication, and erection are so closely interrelated they are impossible to separate, and yet a significant reduction in either fabrication or erection can change the design of the structural system. Bethlehem Steel, in its book *High-Strength Bolting for Structural Joints,* says operating costs are reduced as much as 40 percent when high-strength bolts are used instead of rivets and fabrication tonnage increased as much as 65 percent. When connection angles are bolted to wide-flange beams, the fabricating cost can be reduced over 50 percent when compared with welding. This is true not only for a large run of wide-flange beams but also for the short runs.

Just because welding can be performed at 10 in./min and the rod cost is only $0.20/lb, this does not mean it is economical. Structural estimators usually compute the rod tonnage by multiplying the structural tonnage by 0.5, which is $1/2$ of 1 percent. This would mean a 500-ton structural job would have 5,000 lb of welding rod.

It is not the cost of welding rod or actual time to weld but the fitting time to clamp the clip angles or shear connections into the correct position for tacking that sets the cost of shop welding. The location for tacking of four clip angles is perhaps the most precise and difficult of all fabricating operations. It first requires the fitter to hold a square on the flange of a beam and have the square provide a $1/2$-in. overlap of the clip angle from the end of the wide-flange beam. Not only must there be the $1/2$-in. extension, but the centerline of the first hole must usually be a 3-in. dimension from the outside flange. The fitter is at that time also required to have the connection parallel to the square or at a right angle to the flange. When the connections are being located at the other end, not only must they be square and the first hole 3 in. from the top flange, but also the overall distance to the outside edge of the clip angles on the other end of the beam must be held. This really requires more than two hands and quite a little walking. The beam must be turned over and the clip angles located, tacked, and welded to the other side of the beam. It is not any wonder they are often out of both square and position and have to be burnt off and replaced. Some contractors, and often the government, will often reject a beam after the clip angle has been burnt off.

TIME-STUDY DATA

The fabricating data shown in Fig. L-1 shows the actual time-study and estimating data for the various operations. The fabricator that obtained this data does between $45 million and $50 million a year in business. The process that the fitter goes through to perform operation 7 shows a standard time or 0.60 hours, or 36 minutes, which is faster than the ordinary fabricator would perform the operation. The welding of the four clip angles, operation 8, is done in 0.39 hours, or 23 minutes, plus some material handling time. It is not the approximately 5 lb of welding rod that really costs or the 23 minutes to weld after the tacking, but rather it is the fitting, measuring, positioning, tacking, and turning the beam over that really costs, plus some unknown factors such as well penetration, quality of weld, and the squareness and dimensional accuracy.

The wide flange (shown at the bottom of Fig. L-1, which shows an addition of four holes in the flange) can be loaded on the 790-MBL multiple-punch beam line, punch all holes in web and flange in one setting without layout, and be unloaded in 5 minutes. The clip angles can be assembled and the bolts tightened by hand on one end by pulling both clip angles back on the bolts as far as possible and tightening. This will take care of the $1/2$-in. projection, parallelism, squareness, and the 3-in. dimension from the flange. The two clip angles for the other end may be assembled and hand tightened and then measured for length and squareness and tightened.

The West Side Steel Company fabricated eight wide-flange beams on their beam line in 33 minutes, 50 seconds. This included loading and unloading of the beams on the conveyors. The eight beams were 27 ft 8$1/4$ in. wide

TABLE 1

FABRICATING COST COMPARED

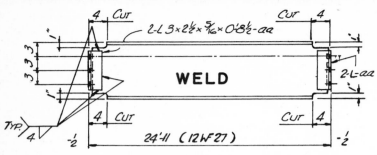

Material Weight   3,463#

BEAMS — WELDED END CONNECTIONS

	Man Hrs.
1. Templets & L.O.	.30
2. Hndle & Cut Bm to Length	.35
3. Cope Flgs	.40
4. Hndle & Shear Conn. Angles, 4 @ .05	.20
5. Punch Conn. Angles, 4 @ .09	.36
6. Material Bay	.15
7. Weld Fit	.60
8. Weld	.39
9. Clean & Paint	.10
	2.85 Mh/Bm

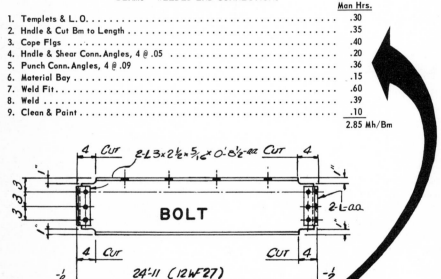

FABRICATION TIME = 5 MINUTES

The wide flange beam can have all (ten) holes punched in both ends and the flange in 5 minutes with accurate dimensions, including loading and unloading. The four connection angles of operations 4 and 5 can now be fabricated in less than two minutes. Operations 4, 5, 7, 8 and part of Operation 1 can all be performed in less than 12 minutes by new methods.

Note: The welding end connection time values were derived at the Metals Estimating Workshop given on November 10-14, 1969 at Wisconsin Center, University of Wisconsin, Langdon & Lake Streets, Madison, Wisconsin by Mr. Jim Holesapple, Chief Engineering, Allied Structural Steel.

FIGURE L-1

Fabricating cost compared. (From Harry Conn, president of W.A. Whitney Corp., Rockford, Ill., *A Realistic Look at Structural Fabricating Costs,* Technical paper No. 550, Metal Fabricating Institute. Used with permission.)

flange 16–36 and averaged 4.25 minutes per beam. Material handling of the beams on and off the beam line required 3.2 minutes and the punching was 1.05 minutes, for a total of 4.25 minutes per beam. The Liebovich Brothers Steel Company of Rockford ran a test and found they were punching in 6 minutes without layout what formerly had required 36 minutes to drill plus layout time.

Some shops have had a standard rule that they would weld-clip angles in their shop providing the wide flange did not require holes other than for the clip angles. This is a costly misconception.

ERECTION COST

Structural-steel erecting firms figure that a bolted-in-the-field design will reduce erection cost about 75 percent over a welded. They figure 1 hour work in the field for each foot of weld. This is because two passes are always needed, and on most jobs three passes are necessary. There is one root pass and two stringer passes required when bolted connections are not used. To do all the welding necessary for the structural connections shown in Fig. L-2 would require over $2^1/2$ hours each for two workers, but it would require only 7 minutes to tighten the nuts and bolts. This is a significant cost reduction.

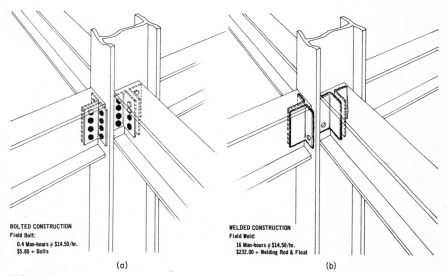

BOLTED CONSTRUCTION
Field Bolt:
  0.4 Man-hours @ $14.50/hr.
  $5.80 + Bolts

WELDED CONSTRUCTION
Field Weld:
  16 Man-hours @ $14.50/hr.
  $232.00 + Welding Rod & Float

(a)   (b)

FIGURE L-2

Cost comparison of (a) field-bolted with (b) field-welded sections. (From Harry Conn, president of W.A. Whitney Corp., Rockford, Ill., *A Realistic Look at Structural Fabricating Costs,* Technical paper No. 550, Metal Fabricating Institute. Used with permission.)

PRODUCTION MULTIPLICATION

Fabricating the same amount of tonnage with 3 workers that formerly required 14 is a significant cost reduction. Fabricating in days the same amount of structural tonnage that formerly required a month is a significant production increase. When production increases from 425 to 1,276 tons with the same size workforce, the efficiency merits recognition.

DESIGN ERROR

The high cost of structural-shape fabricating has been primarily a result of a lack of analysis of designing and detailing and of the proper cost identification of the various fabricating operations such as layout, fitting, riveting, punching, and method of making clip angles.

The high cost starts with a very erroneous and costly note that is put on the structural drawings by the structural engineers: "Weld all shop connections, bolt in field." The note should be: "Bolt all shop connections and bolt in field."

Because of this directive, the detailer is often reluctant to ask for a deviation to allow him to bolt all shop connections, and therefore, perpetuates the costly error of welding shop connections. The structural engineers do not care whether they are bolted or welded, but they should know that welded connections cost over 50 percent more than bolted connections.

Bolted connections have definite advantages over welded connections. Bolted connections are better because there is no doubt about penetration, slag inclusion, and weld quality; they take less skill, are faster to fabricate, and have a lower cost, and because of them, corrections (if needed) are easily made in the field. It takes over 12 months to train a certified welder and 15 minutes to train a person to tighten a nut correctly.

IDENTIFICATION AND QUANTIFICATION

Over 60 percent of structural fabrication cost is in material handling of the shapes for various operations and in layout costs. Layout cost is time-consuming and very expensive and should be eliminated. A welded wide-flange beam has nine operations that include: template layout; handle and beam cutting to length; flange coping; handling and shearing connections or clipping angles; punching connecting angles; moving material to bays; welding, fitting, or tacking; completing welding; and cleaning and painting.

13
CONSTRUCTION PHASE

Value analysis during construction may be initiated by either the owner-designer or the contractor. If the design team has previously made good application of value analysis, the range of reasonable alternative changes remaining at this stage of the project should be substantially limited. As construction proceeds, however, there are inevitably some improvements which either become obvious or are pointed out by the contractor.

Value-analysis changes during construction can be categorized as either internal or external to the construction organization. An internal value-analysis change would be completely within the plans and specifications, requiring no owner or designer approval. The purpose of this type of change is to provide the contractor with an economic edge. The changes in some cases may be identified during the bidding phase, and some of the savings may return to the owner through the contractor submitting a lower bid to assure that he receives the opportunity to perform the work. Prebid value analysis by contractors often includes items with substantial savings. In the postbid phase, the construction value-analysis changes tend to be substitutions of materials or deletions of superfluous items.

An external value-analysis change would be one which requires the concurrence or approval of the owner or his designer, since it affects materials or specified method of installation.

INTERNAL CONTRACTOR VALUE ANALYSIS

Analysis of Techniques

In preparing a construction bid, one contractor noted that a major item of the work involved in the renovation of a large tank would be the cost of developing a system of false work within the tank in order to accomplish major alterations to the roof of the tank. Analyzing the requirements to provide this working level, the contractor determined that the real function was "Provide Platform." Observation of the tank walls indicated that they were sound, and in preparing his bid he submitted an alternative which was based upon filling the tank with water and floating a raftlike staging on the water. The result was a substantial saving, permitting this contractor to win the bid.

In Detroit, contractor Darin and Armstrong bid competitively for a lump-sum project. Their base bid was $10 million, which was in the same range as other bids received. However, Darin and Armstrong had performed a value analysis on the scope of work and submitted certain recommended value-analysis modifications which, if accepted, could reduce the base bid to $8,800,000. The owner accepted, and Darin and Armstrong was assigned the project (Ref. 1).

Martin K. Eby Construction of Wichita, Kansas, had a contract for $9 million to erect two bridges across the Arkansas River at Little Rock (Ref. 2). The contractor came up with a plan for the erection of new bridges which utilized the old bridges for support during construction, saving a substantial amount of money in false work. The new structure was made up of four arches and two continuous spans arranged in pairs, while the old structures were concrete arches. The contractor had to provide a means of removing the old arches after the new bridge was in place. This was achieved by a demolition specialist (Explosives Contracting, Inc., of Little Rock, Arkansas) who developed a plan which would drop the arches straight down into the river below with no damage to the new substructure or excessive throw of debris. The savings made possible by use of the old bridge for false work was, in turn, made possible by creative application of new technology in construction blasting.

The Gordon H. Ball, Inc., construction company modified the anticipated method of installing rail sections with dramatic effect (Ref. 3). The design had envisioned lowering of standard 39-ft sections of rail into an excavation for a station and then welding them into strings on the rail bed. Instead, Ball shop-welds the rails into 507-ft strings. The shop welding is faster, costs less, and is under a higher degree of quality control and thus provides fewer rejects. The 507-ft strings are pulled through the city streets on modified logging dollies and run into the subway on streetcar tracks. The contractor estimated that substantial amounts of money were saved by eliminating 90 percent of the field welding on more than 25,000 ft of rail.

Equipment and Materials

Through analyzing the various procedures which they use and investing in improved equipment or materials, contractors may save substantial amounts of money. In many cases, the investment is minor and response immediate. Figure 13-1 shows the use of a clipping device for the installation of ties for reinforcing bars. This device replaces two previous operations: a loop or wire to tie the cross tie,

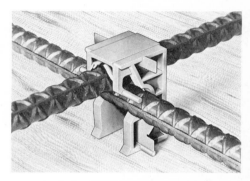

FIGURE 13-1
Clipping device for rebars for concrete slabs. Rebars are tied in place and supported in one operation with the use of G-Loc Clip-Chairs and Guns that automatically lock intersecting bars together. As shown, the steelsetter can use them in a standing position. (Courtesy of Gateway Building Products, Division of Imoco-Gateway Corp., Chicago.)

which does not have the rigidity of the plastic clip, and a chair used to hold the rebar the proper height off the form.

In tunneling work in particular, contractors are investing in major pieces of equipment specifically for the purpose of drilling, and often the equipment is amortized over a single project. Tunneling moles cost a substantial portion of the contract price, but properly applied they save time and money and at the same time produce a superior result.

As reported in *Engineering News-Record* (Ref. 3), the Ball Company utilized a machine made of modified standard equipment in placement of track for the Washington, D.C., metro system. Ball used ganged bushing hammers on the front of an industrial sweeper. The hammers are used to roughen the concrete in the tunnel invert and in the same pass sweep up the chips.

The same company then followed with a modified curb-and-gutter slip form paver to place dual track pads, maintaining a tolerance of plus or minus ⅛ in. on curves and straight sections. For the final step in the track installation, Ball built a custom-made drilling rig to simultaneously drill eight precisely positioned and aligned holes in the pads. The drilling rig also dispensed a measured amount of epoxy in each hole.

The decision to utilize special equipment such as custom-made form work must be analyzed economically on the basis of the number of uses or the production life of the equipment or material. Often, material custom made for one job has limited salvage value for other projects.

Figure 13-2 illustrates the method used to provide mechanized equipment in a restricted space. At Walter Reed Army Medical Center, construction of a $150 million medical complex included renovation of existing space, constant shifting of services, and work in restricted spaces. In this instance, piling had been driven in in a restricted space, and excavation to final grade was going to have to be by expensive hand excavation because a new patient ramp blocked access to the restricted work site. Instead, the contractor's analysis determined that it would be less expensive to rig a front-end loader into the space to do the excavation and pick it out afterwards.

TEAM PLAY

For many reasons, some legal, there is often an inability to perform the construction phase as a true team. In the private sector, there is the opportunity on occasion. In one instance, the use of a construction manager, Carl A. Morse (CAM), was the focus of a combined team ap-

FIGURE 13-2
Portable bulldozer—mechanized equipment in a restricted space. (Pete Esker, Courtesy of *The Service Stripe,* Walter Reed Army Medical Center, Washington, D.C.)

proach to construction. Various approaches were value analyzed with an emphasis upon time and sequencing (Ref. 4).

The project was the $73 million United California Bank (UCB) Los Angeles Headquarters Tower. This building was started in August 1971 with an original projected opening date of March 1975. Based upon the acceleration achieved by the contractor team under the direction of CAM, beneficial occupancy was taken in October 1973. One of the keys was a recommendation on the part of the steel erector Herrick Corporation, which suggested that the structure should be erected from the outside toward the center, even though this is the reverse of the usual sequence. The approach allowed more area to be available to welders and permitted two floors to be erected concurrently. The cycle was two floors per week, cutting erection time by 33 percent.

The construction manager emphasized production schedules, balancing schedules between each trade so that no trade held any other behind. Emphasis was placed upon compression of sequences or shifting sequences to produce faster results.

The construction manager was also able to analyze the potential for use of overtime. One example was the limited space available for building the enclosures for the elevator shafts. The construction manager solved this by having the elevator mechanics working during the day and establishing a night turn and judicious use of overtime for the carpenters.

The construction-management approach provides the vehicle for applying value analysis on a continuing basis throughout construction, but to be fully effective the flexibility of owner support is required. Ultimately, the owner pays in every way on every job, but with the flexibility noted the owner can get more results per dollar invested.

The construction manager is a vital component of the owner organization in construction. Presumably organizations with their own in-house staff would do just as well, but the in-house staff develops the inherent blind spots of a bureaucracy. The construction manager should apply value analysis and may apply it with a staff analyst team or through the use of consultants.

EXTERNAL CHANGES

As noted, external changes are those which require approval of the owner or designer. Changes of this type are initiated either by the owner-designer or by the contractor. Before value analysis and the associated value-engineering change proposal (VECP), most changes relating to design factors were initiated by the owner or the owner's designer.

Owner-Designer Initiated Changes

During the construction, particularly in the early stages, certain areas of cost-saving opportunity may be identified by value analysis. This analysis may be formalized, or it may be intuitively applied. Outside of the federal establishment or the private sector, changes of this type are generally total credits to the owner. In order to avoid an unfair re-scoping of a contract, many lump-sum contracts include an add-deduct factor. Where credits are given, lower unit prices are applied, since in the overall assembly of a lump-sum price the contractor is often required to spread his general conditions and overhead over the work-in-place items. Thus, use of a set unit price to lower the scope of a project would unfairly deduct certain areas of general conditions or overhead which the contractor still had to supply.

In a properly negotiated credit change order, there should be an opportunity for both sides to come out ahead. In many quasi-public and public situations, the owner is not allowed to give the contractor a money credit (with the exception of the add-deduct approach noted above). However, in a properly structured change, there is often a concomitant savings in time to the contractor. Where this occurs, he is able to save on his ultimate overhead through more rapid conclusion of the project.

An instance of this type occurred at the Greater Pittsburgh Airport in a change in surface. The contract called for a textured paint on all plastered ceilings in the corridors of the terminal building. This affected approximately 45,000 sq ft of painted surface.

The job was under pressure to meet schedules, and all work factors were tightly coordinated. Painting was going to cause a definite delay.

The contractor and owner agreed that instead of painting ceilings, plasterers would finish the ceiling with a swirl finish which would be the final surface. The net saving to the owner was $11,500 and a substantial time saving for the general contractor.

In the mobilization stage of the same contract, the general contractor was to provide $150,000 in temporary wood structures to be used as docks for the airlines. These structures would have to be dismantled at the conclusion of the work, with a maximum salvage value of $10,000 offset by the cost of demolition.

The contractor recommended that temporary trailers be employed. The cost factor was essentially the same, but the salvage value was much higher. The contractor had a substantial savings in time and labor, and so both owner and contractor were well ahead through the implementation of this value change.

Unfortunately, without a straight cost incentive, the contractor's general motivation is to reduce the value of the credit change order, since he often interprets the change (sometimes correctly) as deleting more than just material and labor and therefore sees it as an infringement upon his profit and overhead return.

Contractor Change Orders

Even without the structure of a VECP, contractors do observe opportunities for mutual savings and initiate the suggestions. One such change was initiated at the Allegheny County Airport runway extension. This extension passed over a new 950-ft tunnel. The wall structure for the tunnel was designed to be 3 ft wide on the bottom, 5 ft wide at the top, and 25 ft in height. This cross section was the center wall structure and was flanked by a standard wall of retaining wall–type on either side. The center wall was not self-standing until roof beams were in place for lateral support. Accordingly, there was a costly temporary false work structure required for temporary support. The contractor analyzed the supports and determined that it would be less expensive to provide a wall of rectangular cross section which required additional concrete and reinforcing. The contractor made this recommendation to the owner, and it was accepted. The result was a cross section of greater strength, at no additional cost. The contractor determined that in both cost of temporary supports and construction time, he had saved several hundred thousand dollars in the contract.

In another construction situation, design required a heavyweight plastic vapor barrier under a large area of slabs. The contractor experienced a substantial rise in the cost of this material and requested a value change. The contractor offered techniques for compacting of the gravel underdrain and wetting down during placement of the concrete in order to preclude dehydration of the concrete mix. The concrete mix design would also be a low-slump mix so that the concrete mixture would not merge with the gravel drain material. In addition, the contractor offered a credit of almost $50,000.

Unfortunately, the designer who reviewed the requested change for the owner reacted with traditional suspicion of the contractor's intentions. After due deliberation but presentation of no logical reasons, the designer stated, "We designed it that way, and that's the way we want it." Certainly, the contractor in this situation anticipated a substantial savings, which was initially motivated because of the high cost and shortage of the particular material. Nevertheless, the designer refused a $50,000 cash adjustment from which the owner could have benefited. (Value analysis was not included as a specified requirement, on either an optional or mandatory basis.) The inclusion of a value analysis–value engineering approach can help to break down the traditional adversarial role built up between owners and designers on one hand and contractors on the other. This attitude is particularly evident in fixed-price contracts, where the designer is often on the defensive in terms of his design's being complete and correct. Any suggestions regarding changes or substitutions are viewed with considerable concern and/or suspicion. The designer's feeling, often with cause, is "It looks okay now, but what will it cost or lead to in the future?" The designer is on safer ground in this situation to maintain the design, since the owner will blame him, often through legal action, if any problems do arise from a change. Conversely, if a change is made, the designer receives no additional compensation and usually no credit. The owner at the same time is interested in maintaining the fixed price which he had going into the contract but which inevitably is adjusted upward through changes of scope.

VALUE-ENGINEERING CHANGE PROPOSALS (VECPs)

The Department of Defense reported on the contractual aspects of value engineering (Ref.5):

> Former Department of Defense contracting methods did not encourage contractors to submit cost reduction proposals which required contractual action. In fact, the opposite was generally true. A reduction in contract price generally meant a comparable reduction in the contractor's fee or expected profit. Since his reward was derived from these and expected profits, a contractor's reluctance to propose cost reduction action in said circumstances is understandable. Now a positive incentive has been created through the development of the DOD value-engineering contract clauses.

> It is now official DOD policy that value engineering techniques shall be fully utilized by DOD contractors and activities wherever they can profitably be employed (DOD instruction 5010.8).

> The DOD VE contract clauses now invite industry to challenge unrealistic government requirements and specifications and to profit by doing so. These clauses are unlike other contract incentives which reward efficient performance according to the stated terms of the contract. VE contract clauses reward the contractor who proposes acceptable changes to the contract documents which will result in better but lower cost defense products. These changes are mutually advantageous to the government and the contractor because both share the results of the savings. The DOD VE contract clauses encourage entrepreneurship by rewarding contractors equitably for the financial risks they undertake to develop such proposals.

GENERAL SERVICES ADMINISTRATION

VALUE ENGINEERING INCENTIVE CLAUSE
(FIXED PRICE CONTRACT)
SECTION 0015

1. *INTENT AND OBJECTIVES*—This clause applies to any cost reduction proposal (hereinafter referred to as a Value Engineering Change Proposal or VECP) initiated and developed by the Contractor for the purpose of changing any requirement of this contract. This clause does not, however, apply to any such proposal unless it is identified by the Contractor, at the time of its submission to the Government, as a proposal submitted pursuant to this clause.

1.1 VECP's contemplated are those that would result in net savings to the Government by providing either: (1) a decrease in the cost of performance of this contract, or; (2) a reduction in the cost of ownership (hereinafter referred to as collateral costs) of the work provided by this contract, regardless of acquisition costs. VECP's must result in savings without impairing any required functions and characteristics such as service life, reliability, economy of operation, ease of maintenance, standardized features, esthetics, fire protection features and safety features presently required by this contract. However, nothing herein precludes the submittal of VECPs where the Contractor considers that the required functions and characteristics could be combined, reduced or eliminated as being nonessential or excessive for the function served by the work involved.

1.2 A VECP identical to one submitted under any other contract with the Contractor or another Contractor may also be submitted under this contract.

1.3 A proposal to decrease the cost of performing the contract solely or principally by substituting another Subcontractor for one listed by the Contractor in his bid is not a VECP. In considering a VECP which, as an incident thereof, would entail substitution for a listed Subcontractor, maintaining the objective of the Subcontractor listing will be taken into account along with factors cited in paragraph 1.1 above.

2. *SUBCONTRACTOR INCLUSION*—The Contractor shall include the provisions of this clause, with a provision for sharing arrangements that meet or exceed the minimum percentage contained herein, in all first-tier subcontracts in excess of $25,000, and in any other subcontract which, in the judgment of the Contractor, is of such nature as to offer reasonable likelihood of value engineering cost reductions. At the option of the first-tier Subcontractor, this clause may be included in lower tier subcontracts. The Contractor shall encourage submission of VECPs from Subcontractors; however, it is not mandatory that VECPs be submitted nor is it mandatory that the Contractor accept and/or transmit to the Government VECPs proposed by his Subcontractors.

3. *DATA REQUIREMENTS*—As a minimum, the following information shall be submitted by the Contractor with each VECP:

3.1 A description of the difference between the existing contract requirement and the proposed change, and the comparative advantages and disadvantages of each; including justification where function or characteristic of a work item is being reduced;

3.2 Separate detailed cost estimates for both the existing contract requirement and the proposed change, and an estimate of the change in contract price including consideration of the costs of development and implementation of the VECP and the sharing arrangement set forth in this clause;

3.3 An estimate of the effects the VECP would have on collateral costs to the Government, including an estimate of the sharing that the Contractor requests be paid by the Government upon approval of the VECP;

3.4 Architectural, engineering or other analysis, in sufficient detail to identify and describe each requirement of the contract which must be changed if the VECP is accepted, with recommendation as to how to accomplish each such change and its effect on unchanged work;

3.5 A statement of the time by which approval of the VECP must be issued by the Government to obtain the maximum cost reduction during the remainder of this contract, noting any effect on the contract completion time or delivery schedule; and,

3.6 Identification of any previous submission of the VECP including the dates submitted, the agencies involved, the numbers of the Government contracts involved, and the previous actions by the Government, if known.

4. *PROCESSING PROCEDURES*—Six copies of each VECP shall be submitted to the Contracting Officer, or his duly authorized representative. VECPs will be processed expeditiously; however, the Government will not be liable for any delay in acting upon a VECP submitted pursuant to this clause. The Contractor may withdraw, in whole or in part, a VECP not accepted by the Government within the period specified in the VECP. The Government shall not be liable for VECP development cost in the case where a VECP is rejected or withdrawn. The decision of the Contracting Officer as to the acceptance of a VECP under this contract shall be final and shall not be subject to the "Disputes" clause of this contract.

4.1 The Contracting Officer may modify a VECP, with the concurrence of the Contractor, to make it acceptable. If any modification increases or decreases the savings resulting from the VECP, the Contractor's fair share will be determined on the basis of the VECP as modified.

4.2 The Contracting Officer may accept, in whole or in part, a VECP submitted pursuant to this clause by giving the Contractor written notice thereof reciting acceptance under this clause. However, pending issuance of a formal change order or unless otherwise directed, the Contractor shall remain obligated to perform in accordance with the terms of the existing contract.

FIGURE 13-3
VECP form. (General Services Administration form 2653, revised June 1974.)

4.3 An approved VECP shall be finalized through an equitable adjustment in the contract price and time of performance by the issuance of a change order pursuant to the provisions of this clause bearing a notation so stating. Where an approved VECP also involves any other applicable clause of this contract such as "Termination for Convenience of the Government," "Suspension of Work," "Changes," or "Differing Site Conditions" then that clause shall be cited in addition to this clause.

5. *COMPUTATIONS FOR CHANGE IN CONTRACT COST OF PERFORMANCE*—Separate estimates shall be prepared for both the existing contract requirement and the proposed change. Each estimate shall consist of an itemized breakdown of all costs of the Contractor and all Subcontractors' work in sufficient detail to show unit quantities and costs of labor, material, and equipment.

5.1 Contractor development and implementation costs for the VECP shall be included in the estimate for the proposed change. However, these costs will not be allowable if they are otherwise reimbursable as a direct charge under this contract.

5.2 Government costs of processing or implementation of a VECP shall not be included in the estimate.

5.3 If the difference in the estimates indicate a net reduction in contract price, no allowance will be made for overhead, profit and bond. The resultant net reduction in contract cost of performance shall be shared in accordance with the provisions of paragraph 7.2 of this clause; and the contract price shall be reduced by the Government's share of the savings.

5.4 If the difference in the estimates indicate a net increase in contract price, the contract price shall be adjusted pursuant to Clause 23, GSA Form 1139 (General Conditions) or Clause 2, SF 32 (General Provisions); whichever is applicable to this contract.

6. *COMPUTATIONS FOR COLLATERAL COSTS*—Separate estimates shall be prepared for collateral costs of both the existing contract requirement and the proposed change. Each estimate shall consist of an itemized breakdown of all costs and the basis for the data used in the estimate. Cost benefits to the Government include, but are not limited to: reduced costs of operation, maintenance or repair, extended useful service life, increases in useable floor space, and reduction in the requirements for Government furnished property. Increased collateral costs include the converse of such factors. Computation shall be as follows:

6.1 Costs shall be calculated over a 20-year period on a uniform basis for each estimate and shall include Government costs of processing or implementing the VECP.

6.2 If the difference in the estimates as approved by the Government indicate a savings, the Contractor shall divide the resultant amount by 20 to arrive at the average annual net collateral savings. The resultant savings shall be shared in accordance with the provisions of paragraph 7.3 of this clause; and the contract price shall be increased by the Contractor's share of the savings.

6.3 In the event that agreement cannot be reached on the amount of estimated collateral costs, the Contracting Officer shall determine the amount. His decision is final

and is not subject to the provisions of the "Disputes" clause of this contract.

7. *SHARING ARRANGEMENTS*—If a VECP is accepted by the Government, the Contractor is entitled to share in instant contract savings and collateral savings not as alternatives, but rather to the full extent provided for in this clause. For the purposes of sharing under this clause, the term "instant contract" shall not include any changes to or other modifications of this contract, executed subsequent to acceptance of the particular VECP, by which the Government increases the quantity of any item of work or adds any item of work. It shall, however, include any extension of the instant contract through exercise of an option (if any) provided under this contract after acceptance of the VECP.

7.1 When only the prime Contractor is involved, he shall receive 50% and the Government 50% of the net reduction in the cost of performance of this contract.

7.2 When a first-tier Subcontractor is involved, he shall receive a minimum of 30%, the prime Contractor a maximum of 30%, and the Government a fixed 40% of the net reduction in the cost of performance of this contract. Other Subcontractors shall receive a portion of the first-tier Subcontractor savings in accordance with the terms of their contract with the first-tier Subcontractor.

7.3 When collateral savings occur the Contractor shall receive 20% of the average one years net collateral savings.

7.4 The Contractor shall not receive instant savings or collateral savings shares on optional work listed in this contract, until the Government exercises its option to obtain that work.

8. *DATA RESTRICTION RIGHTS*—The Contractor may restrict the Government's right to use any sheet of a VECP or of the supporting data, submitted pursuant to this clause, in accordance with the terms of the following legend if it is marked on each such sheet:

The data furnished pursuant to the Value Engineering Incentive Clause of contract _____ shall not be disclosed outside the Government, or duplicated, used, or disclosed in whole or in part, for any purpose other than to evaluate a VECP submitted under said clause. This restriction does not limit the Government's right to use information contained in this data if it is or has been obtained, or is otherwise available, from the Contractor or from another source, without limitations. If such a proposal is accepted by the Government under said contract after the use of this data in such an evaluation, the Contractor shall have the right to duplicate, use, and disclose any data reasonably necessary to the full utilization of such proposal as accepted, in any manner and for any purpose whatsoever, and have others so do.

In the event of acceptance of a VECP, the Contractor hereby grants to the Government all rights to use, duplicate or disclose, in whole or in part, in any manner and for any purpose whatsoever, and to have or to permit others to do so, data reasonably necessary to fully utilize such proposal on this and any other Government contract.

Ref. 5 notes that in a 4-year period, more than 4,900 contractor VECPs were submitted, of which about 55 percent, or 2,700, were approved. The average contractor's share was 43 percent of the savings, which amounted to more than $50 million in the 4-year period.

Since construction is a significant part of the DOD contractor base, the Naval Facilities Command and the Corps of Engineers were among the first to implement the VECP approach in the construction field.

The VECP contract generated by DOD is in accordance with the Armed Services Procurement Regulation (ASPR) and is rather complex, since it must treat all forms of contracting. Recently, the Government Accounting Office commented favorably on the clarity of the GSA VECP form, which is shown in Fig. 13-3.

Public Building Service Example

In Ref. 6, an example VECP situation is developed. In the hypothetical situation, the contractor has submitted a basic substitution VECP which suggests the use of fiber glass lavatories instead of vitreous china lavatories.

Figure 13-4 is a breakdown of the potential cost savings. The estimating figures indicate that the cost of putting in the contract requirements is $25,155 and that the cost of the proposed approach is $15,827. This results in a net savings of $9,328, of which the government's share would be $3,731.

Note that the contractor probably has not broken out this single item as a contract unit price or pay item. Accordingly, the negotiation figures are open to review. Reference 6 notes that the government must be prudent and careful but not overzealous in evaluating the estimated cost savings. The VECP has the advantage of being essentially self-controlling. The result of the VECP is not net cash increase in the size of the contract but rather a credit deduction in favor of the government. Accordingly, the government's share is a function of the size of the savings.

Reference 6 cites an example wherein a contractor submitted a VECP claiming a $10,000 savings, of which he would get a $5,000 share. The government valuator argued that since it had cost $1,000 to process the successful VECP into a change order, the government was willing to allow savings of only $9,000. Accordingly, the contractor's portion of the savings was $4,500. However, the real effect was to reduce the government's share to $4,500. Negotiation of VECP change orders requires a balanced view rather than the unfortunate adversarial approach which often develops in contractual negotiations.

SUBCONTRACTOR COMPUTATION

ITEM	QUANTITY	UNIT	UNIT PRICE MATERIAL	UNIT PRICE LABOR	TOTAL UNIT PRICE	SUBTOTAL
VITREOUS CHINA LAVATORIES (24 X 20 flat slab)	120	EA	$43.80	$ ---	$43.80	$5256
EXTRA FOR COLOR	24	EA	13.50	---	13.50	324
SETTING FIXTURES	480	MH	---	9.15	9.15	4392
ROUGH-IN MATERIAL	120	EA	18.25	---	18.25	2190
LABOR	720	EA	---	9.15	9.15	6588
				DIRECT COSTS		$18750

PRIME CONTRACTOR COMPUTATION

ITEM	QUANTITY	UNIT	UNIT PRICE MATERIAL	UNIT PRICE LABOR	TOTAL UNIT PRICE	SUBTOTAL
CERAMIC TILE (2'6" ON WALL X 3')	900	SF	$ .65	$ 1.50	$ 2.15	$1935
(2' ON FLOOR X 3')	720	SF	.80	.95	1.75	1260
CERAMIC BASE	360	LF	1.10	1.15	2.25	810
FIXTURE HANGAR	120	EA	7.00	13.00	20.00	2400
				DIRECT COSTS		$ 6405
			TOTAL COST OF CONTRACT REQUIREMENT - - -			$25155

SUBCONTRACTOR COMPUTATION

ITEM	QUANTITY	UNIT	UNIT PRICE MATERIAL	UNIT PRICE LABOR	TOTAL UNIT PRICE	SUBTOTAL
FIBERGLAS LAVATORIES (20 X 18 splash back)	120	EA	$24.20	$ ---	$24.20	$2904
SETTING LAVATORY COUNTER TOPS	420	MH	---	9.15	9.15	3843
ROUGH-IN MATERIAL (BANKS OF THREE)	40	SETS	44.75	---	44.75	1790
LABOR	600	MH	---	9.15	9.15	5490
VECP PREPARATION	4	MH	---	10.00	10.00	40
				DIRECT COSTS		$14067

PRIME CONTRACTOR COMPUTATION

ITEM	QUANTITY	UNIT	UNIT PRICE MATERIAL	UNIT PRICE LABOR	TOTAL UNIT PRICE	SUBTOTAL
LAVATORY VANITY BASE UNITS	120	EA	$10.00	$ ---	$10.00	$1200
INSTALLATION	120	EA	---	4.00	4.00	480
VECP PROCESSING	4	MH	---	10.00	10.00	40
SUBCONTRACT RENEGOTIATION	4	MH	---	10.00	10.00	40
				DIRECT COSTS		$ 1760
			TOTAL COST OF PROPOSED CHANGE - - -			$15827

```
                        SUMMARY

        Contract Requirement Cost        $25,155

        Proposed Change Cost              15,827

              Net Savings                  9,328

              Government Share             X 40%

          Reduction in Contract Price    $ 3,731
```

Estimate of Change in Contract Price

FIGURE 13-4
Estimate of change in contract price—breakdown of potential cost savings. (Public Building Service/GSA Manual P 8000.1, *Value Engineering*.)

UNITED STATES OF AMERICA
GENERAL SERVICES ADMINISTRATION

Public Buildings Service
Washington. D C   20405

General Contractors
Anywhere, U.S.A.

Gentlemen:

IN REPLY REFER TO:
DATE: July 20, 1971
CONTRACT NO: 71-88888
PROJECT TITLE: Any F.O.B.

Your VECP No. 12, dated July 4, 1971, proposing the substitution of fiberglas lavatories for vitreous china lavatories has been carefully evaluated. Unfortunately, it cannot be accepted for the following reasons:

a.   The resinated compound of fiberglas offered has a flash point of 165 degrees F. In the eventuality of the material becoming consumed in a building fire, there is a likelihood that poisonous gases would be released. This would impair safe egress of building occupants and fire fighting.

b.   The product does not have a U.L. rating or certification of compliance with ANSI Code K66.7.

I encourage your continued interest in our value engineering program and thank you for your participation to date.

Sincerely,

Copy to:
    CME/VE
    Contract Files
    Construction Engineer
    Etc.

Keep Freedom in Your Future With U.S. Savings Bonds

FIGURE 13-5
Sample disapproval letter. (Public Building Service/GSA.)

In Fig. 13-5, a sample disapproval letter is shown. The purpose of the government is best served by a positive view toward VECP programs. Accordingly, standing procedures direct that personnel must be careful and courteous in disapprovals and that they must provide specific reasons for the disapproval. The submitter has, then, the opportunity to review his VECP and possibly resubmit for approval.

UNITED STATES OF AMERICA
GENERAL SERVICES ADMINISTRATION

Public Buildings Service
Washington. D C   20405

General Contractor
Anywhere, U.S.A.

IN REPLY REFER TO:
DATE: July 20, 1971
CONTRACT NO.: 71-88888
MODIFICATION NO.: V123
PROJECT TITLE: Any F.O.B.

Gentlemen:

Your VECP No. 12, dated July 4, 1971, proposing the substitution of
fiberglas lavatories for vitreous china lavatories under the above
contract is approved subject to the following conditions:

a.   Lavatories in washrooms for maintenance personnel will not be
      changed. The Government will have the right to select colors
      of the fiberglas lavatories from the manufactures standard stock.
      The Contractor will submit certification of product compliance
      with ASNI Code K66.7.

b.   Approval is based upon a credit of not less than $9,328 to work
      under this contract, subject to upward negotiation, and sharing
      in accordance with the provisions of the VE Incentive Clause of
      this contract.

c.   The Contractor's estimate of collateral impact amounting to a
      one year average savings of $3,900 is deemed reasonable, subject
      to further review and sharing in accordance with the provisions
      of the VE Incentive Clause of this contract.

d.   Approval is based upon no change in contract time for completion.

Accordingly, you are directed to proceed with the change. You will be
notified in the near future regarding final negotiations and consummation
of this approval by formal modification to the contract.

Sincerely,

Copy to:
     CME/VE
     Contract Files
     Contracting Officer
     Construction Engineer
     Architect-Engineer
          Etc.

Keep Freedom in Your Future With U.S. Savings Bonds

FIGURE 13-6
Sample approval letter. (Public Building Service/GSA.)

Figure 13-6 is an approval letter of the example VECP. Note that it guarantees a credit of not less than $9,328 and refers to collateral savings as well. In Fig. 13-7, the collateral figure estimate is developed. Using the recommended approach, the contractor has suggested that the annual costs of maintenance using the fiber glass fixtures will be substantially less, and he suggests that the figure is $3,900. For col-

```
WASH FLOORS -

        10 floors x 2 rooms/floor x 150 SF/room x 52 times/year x
        20 years x $.10/SF/time =                                    $312,000

WASH WALLS -

        10 floors x 2 rooms/floor x 400 SF/room x 2 times/year x
        20 years x $.10/SF/time =                                    $ 32,000

                        COLLATERAL COST OF CONTRACT REQT             $344,000
```

```
WASH FLOORS -

        10 floors x 2 rooms/floor x 114 SF/room x 52 times/year x
        20 years x $.10/SF/time =                                    $237,120

WASH WALLS -

        10 floors x 2 rooms/floor x 355 SF/room x 2 times/year x
        20 years x $.10/SF/time =                                    $ 28,400

GOVERNMENT COSTS -

        Revise drawings for as-builts
        12 sheets x 4 MH/sheet x $10.00/MH =                         $    480

                        COLLATERAL COST OF PROPOSAL                  $266,000
```

```
SUMMARY ESTIMATE

        Collateral Cost of Contract Reqt.      $344,000
        Collateral Cost of Proposal          -  266,000

            20 Year Net Collateral Saving       $ 78,000

            Average One Year Saving             $  3,900

                    Contractor Share               x 20%

            Increase in Contract Price          $     780
```

Summary of Collateral Sharing

FIGURE 13-7
Collateral sharing estimate. (Public Building Service/GSA.)

lateral savings, the contractor gets 20 percent of the first year's savings. This figure is not a credit deduction to the contract but an actual add change. In Fig. 13-8, GSA Form 1137 modifying the contract is shown. The figures are somewhat deceptive, since the contractor's share of the change which he retains is actually $5,597 plus the $780 for collateral savings for a total savings to the contractor of $6,377.

PAGE 1 OF

1. BUILDING NAME AND ADDRESS Any F. O. B.		**GENERAL SERVICES ADMINISTRATION** **PUBLIC BUILDINGS SERVICE**
2. PROJECT NO 36999	3. CONTRACT NO. 71-88888	
4. REQUEST NO N.A.	5. DATE OF REQUEST N.A.	**REQUEST, PROPOSAL, AND ACCEPTANCE** **COVERING CONSTRUCTION CONTRACT MODIFICATION**

6. NAME AND ADDRESS (INCLUDE ZIP CODE)

General Contractors
Anywhere, U.S.A.

REQUEST — Please submit your proposal, signed in quadruplicate, for change(s) outlined below as modification(s) to the above numbered contract.

7 ITEM NUMBER	8. DESCRIPTION OF WORK	9. AMOUNT DEDUCT	ADD
1.	In accordance with the direction to proceed granted on July 20, 1971, substitute fiberglass lavatories for vitreous china lavatories. Net contract savings $9,328 X 40% Government share. Average one year collateral savings $3,900 X 20% Contractor share. Change in Contract price. Item 1 is accepted pursuant to the provisions of the VE Incentive Clause of this Contract.	3731. $2951.	780

10. REQUESTING OFFICIAL N.A.	11. TITLE N.A.

The undersigned contractor agrees to perform any or all of the above changes for the amount indicated. No work on any of the above changes shall be started until this proposal is accepted by the contracting officer.

12 CONTRACTOR'S SIGNATURE AND TITLE	13. CONTRACTOR'S PROPOSAL NO. 12	14. DATE OF PROPOSAL July 4, 1971
15. ITEMS ACCEPTED	16. DATE OF ACCEPTANCE Sept. 10, 1971	17. APPROPRIATION
21. THE UNITED STATES OF AMERICA, BY (CONTRACTING OFFICER)	18. AMOUNT ADD $ DEDUCT $	19. ALLOTMENT NO. 20. CONTRACT CHANGE NO. V123

GSA FORM JUL. 69 **1137**

FIGURE 13-8
Form for contract modifications. (Public building Service/GSA form 1137.)

VECP Motivation

In Ref. 5 it is noted that "top industry management does not always fully understand the intent and objectives of the DOD VECP program, and consequently sometimes fails to give it full support. Aggressive, successful, contractor VECP programs were usually found where top management does fully understand the objectives of the DOD program."

DOD suggests that the VECP program would probably increase if contractors could relate their savings shares to augmentation of income and return on value-engineering investment rather than cost reduction for DOD. One contractor with this view noted that his more than $1 million in profit enhancement as a result of VECPs was equivalent to $20 million in new business.

In Ref. 6, the Public Building Service notes the following reasons most frequently given for lack of contractor motivation in the development of VECPs:

1. Program promotion was not followed through with support at all levels of the procuring activity. Frequently, contractors found that those reviewing their proposals were not sympathetic to the spirit or intent of the program. Some reviewers thought the program was a giveaway without realizing that the act of disapproval of a valid VECP meant the contractor would keep all the money regardless.

2. Review time was often inordinately long. Sometimes contractors did not even receive replies to proposals in which they had invested resources in preparing. VECP review took longer than normal shop drawing review without consideration of the contractor's desiring a decision to maintain his construction schedule.

3. Reasons for disapproval of VECPs were often diplomatic and bureaucratic. Proposals were treated unprofessionally by failure to give valid technical reasons for unacceptability. The tone of many disapproval letters gave credence to contractor opinion that the government spent more effort in figuring out excuses to disapprove ideas than on how they could be made acceptable.

4. And lastly, contractors found that the government would not accept the same degree of risk in approving a VECP that they would take in approving any other change order for extra work or correction of a deficiency. Regardless of dollar value, VECPs were reviewed to death, meticulously analyzed, dissected, and viewed with suspicion much more so than was done for any other type of change.

The PBS finds that contractor training levels are less than they should be, as indicated by such circumstances as high disapproval rates, failure to participate, many marginal-quality ideas, incomplete proposals, and failure to repeat participation. To counter this, the PBS has undertaken a number of supporting activities, including permitting contractors under contract with the government to attend value-engineering training courses for government employees, supporting

Associated General Contractors training efforts, hosting one-day seminars to promote contractor participation, and training conferences for individual contractors when necessary to improve participation.

SUMMARY

The VECP has provided the greatest range of activity in value engineering in construction. It is relatively easy to establish and undertake, requiring a minimum of training on the part of the owner's staff. The contractor may develop his value-engineering proposals utilizing traditional value-engineering value definition and procedures, or he may tend to apply informal techniques to come up with money-saving ideas. In either event, the results operate to the advantage of the government, although certainly the potential for more and larger proposals would be available to those contractors who utilize the proven discipline of the various phases of a full value-analysis application.

Many states, counties, and municipalities which are interested in the value-analysis approach are unable, under existing legislative restraints, to share savings suggested by the contractor. This limits the motivational aspect of the VECP and points out the need for revised legislation and regulations to permit this very obvious saving.

Often, owners are reluctant to share value-engineering savings, failing in a shortsighted outlook to appreciate the fact that there would be nothing to share if the contractor were not willing to come up with the successful proposals and to implement them.

REFERENCES

1. *Building Design & Construction,* May 1974, p. 47.

2. *Contractor's & Engineers Magazine,* Aug. 1973, p. 14.

3. *Engineering News-Record,* Jan. 17, 1974, p. 26.

4. *Engineering News-Record,* Oct. 4, 1973, p. 42.

5. "Principles and Applications of Value Engineering," Department of Defense Joint Course, U.S. Government Printing Office, Washington, D.C., pp. 11, 12, 17, 18.

6. Public Building Service, GSA Handbook P 8000.1, *Value Engineering,* U.S. Government Printing Office, Washington, D.C.

CASE STUDY M
CONSTRUCTION PHASE (PBS VECPs)

The following information was summarized from the *Value Engineering Annual Report—Fiscal Year 1973*. This annual review is prepared by the Public Building Service under the direction of the Director for Value Management Donald E. Parker and Deputy Director Dale E. Daucher.

Table M-1 is a statistical summary of the contractor VECP program for fiscal year 1973.

During fiscal 1973, 63 percent of the VECPs submitted were approved. The total value (government and contractor) of the savings was substantially more than $2½ million.

The Public Building Service makes available information on VECPs approved, disapproved, and withdrawn on the form shown in Fig. M-1. In this VECP, the contractor points out the savings to be achieved by substituting plain-sliced walnut doors rather than quarter-sliced black walnut. The savings (combined) is $97,000, shared equally. The VECP detail report on GSA Form 2698 gives the vital statistics of the change, the contractor, the contract, and the project.

On the same form shown in Fig. M-2 is a disapproved VECP. At the bottom of the form is the reason for disapproval.

In Fig. M-3 is a VECP which was withdrawn because of the press of time.

In the 1973 program, the government realized a return of $13.67 for every dollar invested in managing the program. The 82 successful VECPs approach Pareto relationship in that the 20 percent with the greatest value produced 67 percent of the results.

Table M-1
Statistics of Contractor VECP program for 1973

PUBLIC BUILDINGS SERVICE				FISCAL YEAR 1973	
PROCESSING SUMMARY					
VECPS	NUMBER OF VECPS			ESTIMATED SAVINGS VALUE	
	TOTAL	ON HAND LESS THAN 30 DAYS	ON HAND OVER 30 DAYS	GOVERNMENT SHARE INSTANT CONTRACT	AVERAGE ONE YEAR COLLATERAL SAVING
ON HAND FY 73	3	0	0	N/A	N/A
RECEIVED	130			N/A	N/A
APPROVED	82	58	24	1162.9	1392.0
DISAPPROVED	39	22	17	396.0	448.4
WITHDRAWN	8	5	3	N/A	N/A
ON HAND FY 73	4	4	0	N/A	N/A

VCP DETAIL REPORT				REPORTS CONTROL SYMBOL OA-45
REPORTING ACTIVITY PMCC				DATE 5-22-73
1 VCP DATE 4-27-73	2. DECISION DATE 5-22-73	3. DAYS IN PROCESSING 25	4. DECISION [X] APPROVED ☐ WITHDRAWN ☐ DISAPPROVED	

5. SUBJECT OF CHANGE *(System and component)*

Architectural - woodwork

6. DESCRIPTION OF CHANGE *(Before and after)*

Contract drawings call for all doors to be quarter sliced black walnut. Quarter
sawing produces relatively narrow boards and thus creates a high amount of waste.
As the walnut tree has a relatively small diameter, quarter sawn walnut is in
limited production. The contractor proposes to substitute plain sliced walnut
which is more readily obtainable and will have less effect on ecology. As plain
slicing utilizes more of the flitch, the price is consequently lower.

7. QUANTITIES INVOLVED

8. CONTRACT TITLE. LOCATION Dept. of Labor superstructure, Washington, DC 49918	9. TYPE OF CONTRACT CLAUSE ☐ SUPPLY ☐ CONSTRUCTION ☐ LEASE		
10. CONTRACTOR'S NAME J. W. Bateson Co., Inc.	11. CONTRACT NO. GS-00B-01109	12. CHANGE NO. V 98	

13. INSTANT CONTRACT COSTS/SAVINGS *(Show in thousands of dollars)*

GOVERNMENT SHARE	$ 48.5		CONTRACTOR SHARE	$ 48.5		INSTANT CONTRACT INCREASE	$
OTHER IDENTIFIABLE GOVERNMENT IMPLEMENTING COSTS	$						

14. COLLATERAL SAVINGS *(Show in thousands of dollars)*

AVERAGE ONE YEAR SAVINGS	$		CONTRACTOR SHARE	$		LIFE CYCLE USED *(In years)*	

15. FUTURE ACQUISITION DATA

UNIT PRICE REDUCTION	$		PROBABLE FUTURE PURCHASE *(Quan.)*			ROYALTY PERIOD *(Expiration date)*	

16. DISAPPROVAL/WITHDRAWAL REASONS

GENERAL SERVICES ADMINISTRATION   **GSA** FORM **2698** (REV. 4-75)

FIGURE M-1
VECP approval on detail report form. (Public Building Service/GSA form 2698.)

Of the 82 successful VECPs, 9 involved a major redesign effort on the part
of the contractor. Naturally, this required a substantially larger investment.
However, the returns were great also. The 9 major redesigns accounted for
60 percent of the savings for the year, at an average combined value of
172,156.

<table>
<tr>
<td colspan="4" align="center">VCP DETAIL REPORT</td>
<td>REPORTS CONTROL SYMBOL
OA-45</td>
</tr>
<tr>
<td colspan="4">REPORTING ACTIVITY
5PCCW</td>
<td>DATE
March 21, 1973</td>
</tr>
<tr>
<td>1. VCP DATE

December 22, 1972</td>
<td>2. DECISION DATE

February 27, 1973</td>
<td>3. DAYS IN PROCESSING

67</td>
<td colspan="2">4. DECISION

☐ APPROVED   ☐ WITHDRAWN   ☒ DISAPPROVED</td>
</tr>
</table>

5. SUBJECT OF CHANGE *(System and component)*

Architectural concrete - sand blasting or form liners

6. DESCRIPTION OF CHANGE *(Before and after)*

Use a textured form liner in lieu of sand blasting of concrete to obtain a textured finish.

7. QUANTITIES INVOLVED

8. CONTRACT TITLE, LOCATION
Environmental Control Laboratory
Cincinnati, Ohio   W.O. #50621

9. TYPE OF CONTRACT CLAUSE
☐ SUPPLY   ☐ CONSTRUCTION   ☐ LEASE

10. CONTRACTOR'S NAME
Turner Construction Company

11. CONTRACT NO.
GS-05BCA-0307

12. CHANGE NO.
VCE 5001

13. INSTANT CONTRACT COSTS/SAVINGS *(Show in thousands of dollars)*

GOVERNMENT SHARE	CONTRACTOR SHARE	INSTANT CONTRACT INCREASE
$ 164.00	$ 164.00	$

OTHER IDENTIFIABLE GOVERNMENT IMPLEMENTING COSTS $

14. COLLATERAL SAVINGS *(Show in thousands of dollars)*

AVERAGE ONE YEAR SAVINGS	CONTRACTOR SHARE	LIFE CYCLE USED *(In years)*
$	$	

15. FUTURE ACQUISITION DATA

UNIT PRICE REDUCTION	PROBABLE FUTURE PURCHASE *(Quan.)*	ROYALTY PERIOD *(Expiration date)*
$		

16. DISAPPROVAL/WITHDRAWAL REASONS
The architect objected that the aesthetics of the building would be impaired.   No form liner that was proposed could make him alter his decision.

GENERAL SERVICES ADMINISTRATION   **GSA** FORM **2698** (REV. 4-75)

FIGURE M-2
VECP disapproved. (Public Building Service/GSA.)

The average combined savings for each of the 82 changes was $31,157, but the average for the 73 smaller changes, which involved principally specification of different materials or deletion of work, was $13,774, or only 44 percent of the overall average.

VCP DETAIL REPORT				REPORTS CONTROL SYMBOL OA-45
REPORTING ACTIVITY PMCC				DATE 12-19-72
1. VCP DATE 10-24-72	2. DECISION DATE 12-5-72	3. DAYS IN PROCESSING	4. DECISION ☐ APPROVED ☒ WITHDRAWN ☐ DISAPPROVED	

5. SUBJECT OF CHANGE *(System and component)*

Mechanical/electrical – elevators

6. DESCRIPTION OF CHANGE *(Before and after)*
Contract plans and specifications call for a standard electric elevator to be installed in the building. In that the structure is only two stories plus basement the contractor proposed to install a hydraulic elevator.

7. QUANTITIES INVOLVED

8. CONTRACT TITLE, LOCATION Federal Office Building, Dover, Delaware #070017	9. TYPE OF CONTRACT CLAUSE ☐ SUPPLY ☐ CONSTRUCTION ☐ LEASE
10. CONTRACTOR'S NAME Consolidated Engineering Co., Inc.	11. CONTRACT NO. GS-02B-16717	12. CHANGE NO.

13. INSTANT CONTRACT COSTS/SAVINGS *(Show in thousands of dollars)*

GOVERNMENT SHARE $	CONTRACTOR SHARE $	INSTANT CONTRACT INCREASE $
OTHER IDENTIFIABLE GOVERNMENT IMPLEMENTING COSTS $		

14. COLLATERAL SAVINGS *(Show in thousands of dollars)*

AVERAGE ONE YEAR SAVINGS $	CONTRACTOR SHARE $	LIFE CYCLE USED *(In years)*

15. FUTURE ACQUISITION DATA

UNIT PRICE REDUCTION $	PROBABLE FUTURE PURCHASE *(Quan.)*	ROYALTY PERIOD *(Expiration date)*

16. DISAPPROVAL/WITHDRAWAL REASONS
The proposal was questioned due to the high water table in the area which would lead to leakage problems at the cylinder. The contractor withdrew his proposal as details could not be worked out in the short time allowed by this procurement schedule.

GENERAL SERVICES ADMINISTRATION   **GSA** FORM **2698** (REV. 4-75)

FIGURE M-3
VECP withdrawn. (Public Building Service/GSA.)

Within the category described as substitution, three major items accounted for savings of $325,100. This reduces the average value of the remaining 70 VECPs to $9,720. Nevertheless, where changes can be submitted with relatively little research or effort, the profitability of pursuing this type of value analysis is still excellent on the basis of effort invested.

Table M-2
VECP Performance Effectiveness

Factors	Not applicable	Great benefit	Some effect	No effect	Some disadvantage	Large disadvantage
Esthetics	32	2	6	37	5	
Maintainability	18	3	18	43	3	1
Reliability	21	1	11	47	2	
Performance	13	3	12	52	2	
Time	16	7	27	30	2	
Standardization	36	4	20	21	1	
Quality	18	3	6	52	3	
Simplification	12	8	48	13	1	
Weight	25	2	16	33	6	
Availability	29	5	19	27	2	
Fire protection	30		3	45	4	
Acoustics	34		3	41	4	
Vibration	36		4	40	2	
Safety	31		5	45	1	

SOURCE: Public Building Service GSA.

The PBS report included a number of secondary benefits evolving from the VECPs, as evaluated in Table M-2.

In the following summary extracted from the successful VECPs, the changes are described. The descriptions are necessarily cryptic but may offer some guidance in suggesting areas which could provide good results if value-analysis studies are made. Conversely, the VECP reports describing disapproved work contain many proposals which sound similar to those which were accepted. The final decision is a composite of the actual situation, the content of the drawings and specifications, the location and nature of the project, and the professional opinions of the parties involved.

It is interesting to note that one firm, Hoffman Construction Company of Portland, Oregon, general contractor for federal office buildings in Portland and Seattle, submitted 35 value-engineering proposals, of which 26 were accepted. The value of the combined savings from the accepted Hoffman proposals is approximately 63 percent of the total savings listed for fiscal year 1973.

1.0 Site Preparation

1.1 Specifications called for existing timber and scrub to be felled, collected, sold, or otherwise utilized and the balance mulched and stored on site. The contractor obtained a valid permit to collect and burn the timber and scrub under controlled conditions. ($15,400)

1.2 Contract drawings required a series of 5-in. holes 5 ft deep, 5 ft on centers to be drilled in the soil and lime injected into these holes under pressure. The contractor used a system of hydraulically injected probes to inject a lime slurry to accomplish the lime stabilization. ($4,000)

2.0 Piling and Footings

2.1 A change was made from driven "H" piles to concrete-filled driven steel-pipe piles. ($35,600)

2.2 Specifications called for 12-in. precast-concrete piles driven to specified resistance. The contractor provided a 12¾-in. steel-tube cast-in-place pile. The tubes had 0.219-in. wall thickness. The piles were reinforced and filled in accordance with the Virginia Department of Highways specifications for this type of concrete pile. Saving estimated at $0.52 per linear foot of driven pile based on bid quantity of 13,400. ($7,000)

2.3 The contractor raised the pile-cap elevation in certain areas to reduce excavation, backfill, form, and concrete. This did not affect structural soundness or future building usage. ($2,600)

2.4 The contractor raised the bottom elevation of certain footings. (A/E investigated and recommended acceptance, provided that a minimum of 6 ft/0-in. coverage was maintained.) Bearing was sufficient at the proposed elevation. ($4,000)

3.0 Foundation Redesign

3.1 The tie-back shoring and the garage exhaust collection header were relocated. The contractor moved tie-back shoring to utilize face of lagging as back form for foundation walls, changed dampproofing to utilize Anti-Hydro in foundation wall concrete, moved garage exhaust collection header into basement area, and eliminated footing at top of caissons. ($43,000)

3.2 The contractor revised the garage foundation by founding the structure on high-capacity (100 tons) bearing piles using 12 in. BP 53 and replaced air tunnels with concrete pipe using rubber-gasketed joints. ($67,800)

4.0 Foundation Drainage

4.1 Specifications called for tile drain piping with open joints covered with brass screen. The contractor installed porous concrete pipe with interlocking tongue and groove joints. ($1,500)

4.2 The contractor substituted corrugated polyvinyl chloride (PVC) pipe for the perforated tile specified for foundation drains. ($600)

4.3 A drainage fill material which was available locally was used in lieu of the specified material which would have required an 80-mile haul. (The local source was available all winter, while specified material could not be obtained until the spring.) ($12,800)

5.0 Foundation Waterproofing

5.1 The requirement to waterproof the pile caps under the courts building portion of the project was eliminated. The slab waterproofing was extended to protect the column base plates. ($10,000)

6.0 Site Piping

6.1 Specifications required concrete encasement of piping between manholes. The contractor provided asphalt-coated steel with fiber glass cloth and calcium silicate insulation, with 3-in. insulation on the steam line and 1½-in. insulation on the return, packaged with anchors, seals and casings complete. ($33,000)

6.2 Contract drawings called for manholes B, C, and D to have a floor drain leading to a 4-in. line which drained to manhole A. At manhole A a steam-operated sump pump raised the drainage to a higher-elevation drain line. The contractor provided each manhole with a separate sump pump. The 4-in. connecting line was installed at a higher elevation, eliminating costly excavation and backfill. ($12,000)

6.3 The same contractor repeated the same detail on another contract. ($11,000)

6.4 The water supply lines for building were changed from cast iron to transite pipe. ($1,000)

7.0 Structural System

7.1 The contractor redesigned and constructed a parking garage using design live load of 50 lb/sq ft and precast sections. The parking garage construction required poured concrete of 100 lb/sq ft live load. The contractor's proposal called for redesigning the parking garage by utilizing precast-concrete sections at a reduced live load of 50 lb/sq ft in lieu of 100 lb/sq ft. ($114,000)

7.2 A precast clad steel frame building was provided in lieu of an architectural reinforced-concrete building. The building was 17 stories with a perimeter of 120 × 150 ft. Columns around the perimeter were 10 ft c-c; with the new design they were 40 ft c-c one way and 30 ft c-c the other way. The floor structure was changed from precast concrete to metal decking with concrete fill. The new framing system permitted the structural height of each floor to be reduced by 6 in. ($260,400)

7.3 The contractor modified the earthquake-resistant metal stud partition connections to overhead steel beam and metal deck. The top of the metal studs were free in deep runner track with ⅜-in. gap between top of studs and runner track to allow movement of partition in lieu of continuous 4 × 4 in. angle with 4 × 4 in. angle spaced sections with slotted holes to allow movement. ($31,000)

7.4 Reinforced CMU was substituted for the reinforced concrete walls above the first-floor elevation. ($32,000)

7.5 The contractor eliminated construction of concrete slab as per detail and substituted a saw cut, or "quick joint", without joint sealant. ($5,200)

7.6 The painting of all structural steel was eliminated, with the exception of the perimeter steel members at the window wall. ($55,800)

7.7 The contractor did away with painting requirements for structural steel encased by spray-on fireproofing or insulation. ($17,700)

7.8 The prime coat of paint on structural steel items to receive sprayed fire protection was eliminated. ($6,100)

8.0 Concrete Materials

8.1 Concrete—Design mix: The contractor furnished alternative 4,000 lb/sq in. warm-tone architectural concrete mix design based on 6.0 bags of cement in lieu of 6.5 bags mix specified. ($3,000)

8.2 100 lb/cu ft sand concrete was used in lieu of 90 lb/cu ft solite concrete for floor fill on all except the third floor. The concrete was still 2,500 lb/sq in. as required but used less expensive fine aggregate (regular sand versus solite sand). ($50,000)

8.3 Contract drawings required gravel fill and polyethelene sheeting under slab areas on grade. The contractor used sand fill for this purpose, which provided a more stable base. ($5,460)

9.0 Metal Decks

9.1 The contractor modified construction from cellular decking to blended deck in all areas specifying metal deck. ($120,000)

9.2 There was a change from wood-formed concrete decks to Q-Lock-99-20-gauge metal deck except areas for which metal deck was specified and other areas which included mechanical floors and computer areas. ($20,200)

9.3 The contractor installed 22-gauge steel roof deck in lieu of the 20-gauge steel roof deck as sufficient strength. ($4,600)

10.0 Fireproofing

10.1 The contract plans showed steel beams and columns to be fireproofed with 2 in. of concrete. At the basement level, a fire-rated ceiling was to be run between the fireproofed beams. The contractor proposed to:

a. Change the column fireproofing to 3-hour rated gypsum perlite plaster.

b. Provide a concrete wainscot on columns in the traffic areas.

c. Lower the fire-rated ceiling beneath the beams, eliminating concrete; the alternative to provide a sprinkler system was accepted by the government, thus providing complete fire protection of structure and contents. ($18,200)

10.2 In lieu of 2-in. textured ceiling board, 1½-in. Industrial Insulation Board was substituted at underside of main sloping roof; concrete fire-proofing was eliminated on all sloping roof beams, and 3-hour boxed vermiculite plaster was used instead; [3-hour sprayed fireproofing en-closed by furring of metal lath and cement plaster was substituted for all interior precast-concrete column covers and 8 WF horizontal beams; and 3-hour rated construction of furring, metal lath, and vermi-culite plaster was substituted for concrete encasement of all columns supporting the main sloping roof as well as the perimeter roof beam for the machine room.] ($15,200)

10.3 The requirement for fireproofing material application for main-building second-floor slab decking was eliminated. ($6,200)

11.0 Exterior Walls

11.1 The contractor utilized the type of surface aggregate for use in the type II precast-concrete panels; approval was given for the Wyoming white aggregate. ($33,200)

11.2 A 1-in. thick insulated glass unit with two ¼-in. plates and a ½-in. air space between was used in lieu of a ⅜-in. laminated reflective glass. Advantages of insulated glass are the ability to withstand greater wind load, a greater insulating value, which will reduce the heating and cooling load of the building, and a 20-year manufacturer's warranty over a 5-year warranty. ($12,700)

11.3 Contract plans and specifications called for concrete beams and col-umns faced with limestone to be dampproofed. Dampproofing would protect the exposed stone from concrete salts leeching through the mortar setting bed and staining the face of the stone. Setting specs called for the limestone to be set with a 1½-in. air space between the concrete and the stone and the joints sealed with elastomeric com-pounds instead of mortar. Thus no capillary action could occur. ($11,700)

12.0 Windows and Window Installation

12.1 Contract drawings showed that a reinforced brick soffit was to be in-stalled above each window. The contractor substituted a soffit made of colored stucco on metal lath instead of brick. Specifications also re-quired that high bond mortar be used; the contractor substituted type S colored mortar. ($108,100)

12.2 The contractor substituted 360-degree single-glazed windows and mo-dified framing with automatic lock feature at the 180-degree closed po-sition in lieu of the windows specified. ($100,000 plus collateral share)

12.3 Continuous $3 \times 2 \times \frac{1}{8}$ in. galvanized tube sill angles were embedded in all precast-concrete window-unit panels where a sill embedment had been required. ($19,000)

12.4 The installation of aluminum siding over existing exterior windows and openings at rear of building was eliminated, and stucco on metal lath with proper provisions for waterproofing was installed around the perimeter of all windows and openings. ($5,342)

13.0 Roofing and Flashing

13.1 Contract drawings called for the roof coping to be fabricated from steel with a painted finish. The contractor installed a combined gravel stop/fascia fabricated from anodized aluminum. ($20,000)

13.2 Copper was used instead of the specified anodized aluminum for fascia and equipment screen on top of roof. ($4,800)

13.3 Roof construction was changed from lightweight insulation fill to rigid board insulation. This entailed the installation of both a minimum 1-hour fire-rated ceiling and 2¾-in. thick "FESCO" board insulation in order to retain the specified coefficient of heat transmission. ($19,200)

14.0 Partitions

14.1 Self-furring metal lath on masonry and concrete walls was eliminated; it was not needed because sufficient bond was generated without the lath. However, lath was retained over joints, at all corners and around openings. ($50,400)

14.2 Drywall with tape and float finish was used in lieu of veneer plaster and gypsum plaster finish on areas as specified. The old system called for ⅜-in. gypsum board with $1/16$-in. veneer coat, the new system for ⅝-in. fire-rated gypsum board for free-standing partitions and ½-in. fire-rated gypsum board for furred columns and walls. ($32,000)

14.3 6-in. metal studs and double ½-in. thick gypsum board on each side were eliminated, and 4-in. metal studs and single ½-in. thick gypsum board on each side were used instead. ($19,000)

15.0 Ceilings

15.1 The perimeter ceiling areas were modified by the elimination of the perimeter gypsum board soffit and recessed blind pockets, and the installation in their place of a continuation of the suspended acoustical tile ceilings directly over to the horizontal sound deadened blind pockets, with recessed track to accommodate vertical blinds. (576 lineal ft of perimeter wall per floor × 34 floors = 19,584 lineal ft.) ($15,800)

15.2 12-gauge galvanized annealed hanger wire for type 1 acoustical unit areas was substituted for stainless steel or nickel-copper alloy wire as specified. ($9,000)

15.3 8-gauge galvanized annealed hanger wire for type 2 acoustical unit areas was used instead of stainless steel or nickel-copper alloy wire as specified. ($400)

16.0 Wall Treatment

16.1 Contract drawings called for marble of varying thickness ranging from 1½ to 5 in. Drawings were silent as to attachment details and conflicted in some instances as to dimension. In addition, specifications required liners to be used as reinforcing (additional pieces of stone of same thickness laminated to back), but there was insufficient room between stone and back-up material. The contractor used ⅞-in. marble in all areas where possible. This saved money without sacrificing function and will eliminate possible future change orders resulting from poor design. ($20,200)

16.2 Gypsum plaster backing under ceramic tile walls was used in lieu of Portland cement plaster backing. Portland cement plaster is the traditional base for ceramic tile because it is relatively waterproof. However, these rooms are not subject to constant wetting but merely maintenance washing. ($38,700)

17.0 Floor Finishes

17.1 Contract drawings showed precast paving panels as 2 ft 5½ in. square and those on another plaza as 4 ft 10⅝ in. square. The contractor proposed using the same larger size for both areas, which would simplify forming and installation. ($6,000)

17.2 The contractor eliminated the use of 24 × 24 in. terazzo squares and substituted 6 × 6 × ½ in. quarry tile in a bed of cement grout with chemical-resistant mortar to resist salts. ($600)

17.3 Contract drawings showed cast-in-place terrazzo with 4-ft cast-in-place terrazzo base to be installed in elevator lobbies and in vending areas. The contractor originally proposed that precast terrazzo tile and regular premolded cove base be provided in lieu of the cast in place. The precast tile contains real marble chips cast in an epoxy matrix which is factory ground, polished, and sealed. Upon review, the use of terrazzo in vending areas was questioned, and it was decided to substitute regular resilient tile in these areas. The contractor's proposal was revised to include this, and the result was greater credit. ($12,700)

17.4 Nonslip stair nosings specified for the pan-type stair treads and landings at stairs were eliminated and replaced by a nonslip finish to the entire tread and landing areas with mason tread grooves on each tread. ($8,800)

17.5 The specifications required that wool carpeting furnished by the contractor be installed in several of the courtroom areas. Most carpet now furnished for these areas by the government is usually an Acrylan acrylic carpet. Samples of such carpet submitted by contractor were acceptable to the judge and to building management. ($2,200)

17.6 The contractor proposed the material for the carpeting be acrylic instead of wool as specified. ($9,000)

18.0 Furnishings and Hardware

18.1 Telephone shelves were installed by the telephone company at no cost to the government rather than by the construction contractor. ($400)

18.2 Contract specifications required highly polished finish on door butt hinges. The contractor used standard polished hinges, which was more in line with other buildings in that area. ($680)

18.3 Contract drawings called for all doors to be quarter-sliced black walnut. Quarter sawing produces relatively narrow boards and thus creates a high amount of waste. Because the walnut tree has a relatively small diameter, quarter-sawn walnut is in limited production. The contractor proposed to substitute plain-sliced walnut, which is more readily obtainable. As plain slicing utilizes more of the flitch, the price is lower. ($97,000)

18.4 Vinyl-clad steel for the air enclosure cabinets was used instead of the specified anodized aluminum. ($13,000)

18.5 The contractor changed from nonstandardized elevator buttons and lanterns to standard or semistandard buttons and lanterns. ($12,800)

19.0 Job Conditions

19.1 Contract drawings showed the existing main lobby to be renovated under that contract. In the time between the preparation of the drawings and the award of the contract, the lobby was completely redone by the occupying agency. Because the existing work was still in excellent condition, it was agreed to leave it as it was and delete the renovation from the new contract. ($14,300)

19.2 The in-place piping and sprinkler heads for five sprinkler systems were reused in lieu of installing new but identical equipment. ($2,400)

20.0 HVAC

20.1 The contractor modified the perimeter heating, ventilation, and air conditioning systems above the fifth floor in order to utilize an all-air system in lieu of a four-pipe system as specified. ($300,000)

20.2 Window unit enclosures to all-air system requirements were substituted for fan coil unit enclosures as specified (contingent on 20.1). ($50,000)

20.3 The contractor substituted an all-air-type (variable-air-volume) heating, ventilating, and air conditioning system in lieu of an air-water induction system above the first floor as was originally designed in order to furnish all perimeter convector cabinets instead of induction unit cabinets and to eliminate the stair smoke vestibule and rotate the stairs 90 degrees, incorporating the vestibule space into the return air shafts. These modifications encompassed 16 floors of the tower building (interior dimensions 122 × 150 ft). ($138,000)

20.4 First floor (20.3) was specified to be served from separate multizone unit supplying through overhead distribution system with hot-water radiation at floor level. The basement and ground floors were to be served by a variable-volume system using induction-type terminal units with hot-water reheat coils. Street steam was substituted as the main heating source in lieu of the installation of steam boilers for heating. The exhaust air tunnel to the garage was eliminated and concrete pipe using rubber gaskets was installed. ($75,800)

20.5 Flame-resistant polyethylene (Dekoron FR) tubing was used instead of copper tubing for certain pneumatic temperature-control line. ($19,700)

20.6 Insulation from return air ducts was eliminated. ($8,300)

20.7 Contract drawings called for a unitary air conditioner in each ante-room with ductwork leading to a rooftop discharge. The contractor substituted a standard package air conditioner installed in the ante-room, with the condenser on a low roof outside. The discharge duct-work was eliminated. ($1,000)

21.0 Mechanical Equipment

21.1 The condensate duplex pump set and the condensate cooler tank were eliminated, and a heating coil was added to the condensate receiver, sized to satisfy requirements as specified for heating coil which was deleted. ($2,200)

21.2 The size of the air compressor was reduced from 66 to 15 cu ft/min for the laboratory building. The building needs for this area were reduced because of tenant requirements. ($1,240)

22.0 Plumbing

22.1 The design specified installation of a complete plumbing system con-sisting of domestic hot and cold water, sanitary sewer, roof and storm drainage, and connections. In place of this the contractor, utilizing plumbing fixtures and other equipment all in accordance with the re-quirements of the National Plumbing Code, used a centralized riser system. ($35,000)

23.0 Electric Power Distribution

23.1 The type of power-distribution system specified was changed to that commonly accepted as standard for commercial office buildings. ($110,000)

23.2 Miscellaneous electrical changes:

   a. All rigid conduits were changed to EMT in sizes up to 4 in., but no EMT was installed underground or in slabs on grade.

   b. Standard National Electric Code approved type supports and hangers gen-erally used in commercial office buildings were installed.

c. Raceway sizes were changed to conform with the conductor fill specified in the 1968 National Electric Code. ($40,200)

23.3 Contract drawings showed that the bus duct was to be run across the ceiling slab of the garage. Where the new elevator lobby occurs the bus duct would terminate in tap boxes and run above the suspended ceiling in conduit. The VECP substituted interblock cable which is supported on cable trays and does not require tap boxes at the suspended ceiling. ($6,800)

23.4 The contractor originally proposed that the transformer secondary bus be changed from copper to aluminum, but later this proposal was revised to running the secondary in an underground concrete-encased conduit in lieu of a busway. ($1,800)

24.5 Stress cone connections for electrical feeder lines were installed instead of the specified pothead connections. ($600)

24.0 Electrical Equipment

24.1 The perimeter telephone outlets and electrical receptacles on all floors from the sixth through the thirty-fifth were eliminated. ($50,000)

24.2 Telephone equipment room conduits were replaced by a cable tray for conductors in lieu of separate telephone conduits shown on the contract drawings. ($2,014)

24.3 The light fixture recess specified was eliminated and replaced by a framing system which provided a simpler framing system, thus permitting easier installation of light fixtures plus the added advantage of better fixture heat dissipation above the ceiling. ($1,590)

CASE STUDY N:
VALUE ANALYSIS IN HEAVY CONSTRUCTION

by Maj. Gen. William L. Starnes, U.S.A. (Ret.)

Author's note: General Starnes is Vice President for Administration of U.S.A.A. Insurance, San Antonio, Texas. His military career of almost 30 years included assignments as the district engineer of the Cape Kennedy District during the final stages of the Apollo launch complex construction and assignment as the deputy director for the building program for military construction performed by the U.S. Army Corps of Engineers. During his last several years with the Corps, Maj. Gen. Starnes was the division engineer for the Ohio River Division. In that assignment, he was responsible for approximately $200 million in new civil work construction programs each year. The Ohio River Division (ORD) is located as shown in Fig. N-1. The division has four districts covering parts of 14 states in the Ohio River drainage basin.

The Ohio River Division has emphasized the application of good management principles with special consideration to the efficient reduction of costs. Our value-engineering program was a highly satisfactory means of controlling construction costs without compromising the intended performance of our projects when completed.

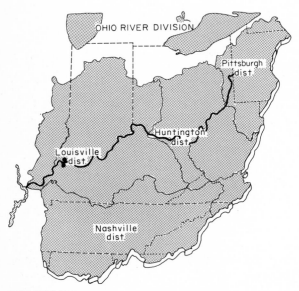

FIGURE N-1
Location of the Ohio River Division. (Courtesy of the U.S. Army Corps of Engineers.)

Twenty to twenty-five years ago, the Corps of Engineers civil works designs were developed at a rather leisurely pace. Each feature was subjected to a rigid comparative analysis to obtain the best project for the least cost. However, under an ever-increasing design work load, there has not been the time available for this extended economical analytical review process. The ever-increasing design work load compounded by added ecological reviews and consideration, coupled with an acceleration of deadlines for new construction starts, and impacted by a systematic decrease in the workforce to perform design has resulted in designs that are safe and aesthetically pleasing—but not always economical.

The Ohio River Division and the Corps of Engineers in general have had to depend more and more upon the value-engineering program to review designs and to identify and refine those features which appear to be too costly.

In the Ohio River Division, under my direction, the validated savings under the value-engineering program increased from $2,372,000 in fiscal year 1970 to over $22 million in fiscal year 1972—and this figure applies to civil works alone. This represents an increase of almost 1,000 percent in a 4-year period. These figures include only validated savings that are initially creditable to the value-engineering program. Many of the design improvements which were identified and developed under the value-engineering program have been incorporated into subsequent designs and have produced a cascade effect of additional savings which cannot be credited to the value-engineering program. Nevertheless, these additional savings have been highly beneficial to the overall construction program.

R. D. BAILEY

R. D. Bailey is a flood control reservoir located in West Virginia. The project is a multipurpose one, constructed for flood control, water quality, and recreation. Figure N-2 shows the downstream view of the outlet portal of the outlet tunnel. At the left is the location of the access road, the overlook, and the operations area. Dam construction had not been started at this time, although it was under design. Construction was planned for later in the year.

The location for the access road in the original design prepared by the Huntington District was along the north side of the hill from the point on U.S. Route 52 to the left abutment of the dam. (Figure N-2 shows the river flowing from upper right to lower left.) This location of the access road required several rock cuts averaging approximately 70 ft in depth. A service road was to be conducted across the top of the dam to service the dam tender's quarters, and operations and maintenance areas. The road went on to the intake structure. The operations and maintenance areas, including the dam tender's quarters, were to be located on the right abutment, just upstream from the center of the dam. Recreation facilities for public use were limited to an overlook that would accommodate only 20 cars and 2 buses.

As a result of a value-engineering study, the alignment of the access road

FIGURE N-2
Ohio River, flowing from upper right to lower left. (Courtesy of the U.S. Army Corps of Engineers.)

FIGURE N-3
Expansion of the recreation and overlook area.

was changed to the south side of the hill. This shift had the effect of utilizing the sun to help reduce snow accumulation on the access road. The new location used an abandoned strip mine. The shift also made possible the expansion of the recreation and overlook area into a broader development, as shown in Fig. N-3. The value-engineering study resulted in a reduction of more than $700,000 in the initial construction costs. Perhaps much more important, and not quantifiable into dollars, the shift converted a horrible strip-mine scar on the mountainside into an attractive and useful facility.

ALUM CREEK DAM

This embankment dam included a gated spillway shown in Fig. N-4. The design for the embankment included a 10-ft wide inclined drain leading to a 3-ft thick blanket drain.

Inclined drains are usually designed to be constructed from local granular materials, which are usually relatively inexpensive.

There are two criteria for sizing the drains: (1) the cross-sectional size to accommodate seepage requirements and (2) a minimum size to accommodate construction equipment. In most cases, the minimal access size for construction equipment is larger than the seepage requirement.

At Alum Creek dam, the contractor's bid price for "embankment, filter drainage layers, and transitional zones" was $7/cu yd. The controlling factor

FIGURE N-4
Gated spillway—Alum Creek dam.

determining this relatively high unit price was the long haul distance from the nearest adequate source of granular material.

Value-engineering study groups singled out this drain and determined that theoretical requirements for seepage for the inclined drain could be satisfied with a drain slightly over 2 ft thick. The group developed a new placement method to reduce the width of the inclined drain from 10 to 5 ft, which would guarantee the necessary theoretical feet. It was determined that a 3-ft layer of impervious material would be placed and compacted, and that a 5-ft wide trench would be excavated through the impervious fill and backfilled with drain materials. An additional 3-ft layer of impervious materials would be placed in a 5-ft wide trench excavated with an offset of 1½ ft from the excavation in the lower left. The excavation would extend a few inches into the lower granular fill and then be backfilled with granular material. This procedure would be repeated in stairstep fashion until the dam was completed. A comparison between the initial and revised approaches is shown in Fig. N-5a and b.

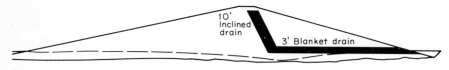

Typical embankment section
Alum Creek Dam

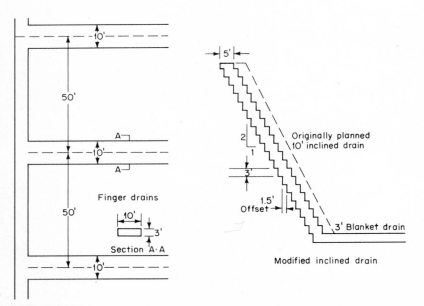

FIGURE N-5
Comparison between (a) initial and (b) revised approaches—Alum Creek dam.

FIGURE N-6
Outlet works under construction—East Fork dam.

The Huntington District value-engineering study team was still not satisfied that the full potential had been developed. The team next designed finger drains 10 ft wide and 3 ft deep on 50-ft centers to replace the blanket drain for the purpose of conducting the incline drain seepage to the embankment toe. Calculations showed that a reduced drain area would be adequate for the job.

Because the placing of the impervious drain material and more precise placement requirements would increase the unit cost, the estimated unit price was increased to $11/cu yd. However, the use of less of this higher-priced material resulted in a net change to the contract, which saved the government $702,900.

EAST FORK DAM

The East Fork dam is a multipurpose dam located on the little Miami River in the Louisville District, Ohio. It is an earth and rock–filled dam and saddle dam. Figure N-6 shows the outlet works under construction.

The preliminary design included a small spillway to be cut in the right side of the dam (Fig. N-7). However, emergency discharges over this spillway would provide a potentially dangerous erosion problem at the downstream toe of the dam, which in turn would necessitate extremely expensive protective features. This potential condition can be noted in Fig. N-7. The value-

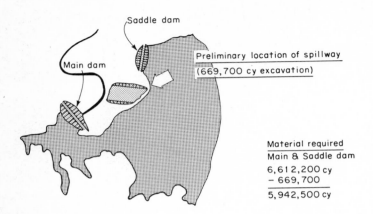

EAST FORK LAKE
East Fork Little Miami River, Ohio

Saddle dam

Main dam

Preliminary location of spillway
(669,700 cy excavation)

Material required
Main & Saddle dam
6,612,200 cy
− 669,700
5,942,500 cy

FIGURE N-7
Small spillway to be cut in right side of dam—East Fork dam
preliminary design.

engineering team located an alternative location for the spillway on the
other side of the dam, as shown in Fig. N-8.

A revised design was developed to provide discharge capacity equal to the
original spillway, thus eliminating the need for expensive protection for the
downstream toe of the dam.

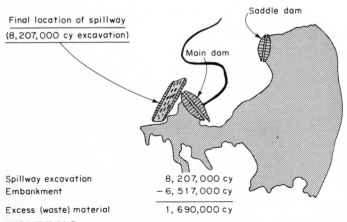

EAST FORK LAKE

East Fork Little Miami River, Ohio

Final location of spillway
(8,207,000 cy excavation)

Saddle dam

Main dam

Spillway excavation	8,207,000 cy
Embankment	− 6,517,000 cy
Excess (waste) material	1,690,000 cy

FIGURE N-8
Alternative location for spillway—East Fork dam.

FIGURE N-9
East Fork dam depth increased by 1 ft.

The latest design, however, resulted in more than 8 million yd of excess material. The value-engineering team then directed its attention to this excess material.

Additional hydrologic routings were made, and it was found that by increasing the height of the dam 1 ft, flood storage would be increased and part of the excess material could be effectively utilized (Fig. N-9). At the same time, the width of the spillway was made narrower, producing a lesser amount of excavation (Fig. N-10).

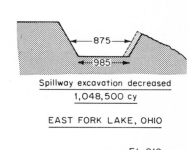

Spillway excavation decreased
1,048,500 cy

EAST FORK LAKE, OHIO

FIGURE N-10
Width of spillway made narrower—East Fork dam.

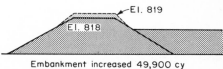

Embankment increased 49,900 cy

The overall results of the value-engineering study were a balanced design and a reduction in cost of $1,138,000. The following is summary of savings.

Decreased spillway excavation (1,048,500 cu yd)	$ 744,000
Less cost of added embankment to raise dams 1 ft (49,900 cu yd)	−11,000
Savings	733,000
Collateral savings	23,000
Other savings:	
Less concrete in spillway control structure	
Other changes to foundation	
Grouting and seepage control	
Transition zone reduced cost	382,000
Grand total savings	$1,138,000

BROOKVILLE FLOOD CONTROL PROJECT

The Brookville Flood Control Project is located in Indiana about 50 miles from Cincinnati. The Brookville dam is on the east fork of the Whitewater River in the Louisville District.

The design provided for the use of all suitable material excavated from the spillway to be placed in the dam embankment. Material in the spillway will withstand erosive velocities up to 10 ft/sec (Fig. N-11). Sections where this

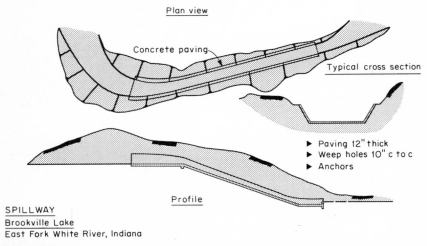

Plan view

Concrete paving

Typical cross section

▶ Paving 12" thick
▶ Weep holes 10" c to c
▶ Anchors

Profile

SPILLWAY
Brookville Lake
East Fork White River, Indiana

FIGURE N-11
Spillway—Brookville dam.

velocity is equaled or exceeded would be protected by concrete paving. Since concrete paving is considerably more expensive than embankment material, the value-engineering study team looked for means to reduce its quantity.

Figure N-12 shows the profile of the spillway area before and after the studies. The paving extends approximately 1,000 ft upstream of the control section with a width varying from 70 ft at the control point to 134 ft at the entrance. In order to delay the point at which flow reached 10 ft/sec, an overexcavation of the approach section was proposed. This permitted the length of paving to be reduced 200 ft, as shown in the "after" section of the Fig. N-12.

Figure N-13 shows Brookville dam during construction. On the left part of the picture, the white area is concrete being placed for the spillway. The deepened excavation area for the approach site can be seen.

The overall net savings of this change were $90,221 and appear in summary as follows:

Over excavation of spillway (25,306 cu yd) and drainage pipe	added	$47,800
Borrow excavation (25,306 cu yd)	deleted	12,600
Concrete paving (1,252 cu yd)	deleted	89,800
Drill and grout anchor bars (1,017 each)	deleted	31,500
Portland cement (1,565 barrels)	deleted	7,600
Cost of value-engineering study and contract modification	added	3,479
Net savings		$90,221

PRIMARY MISSION: NAVIGATION

The major mission of the Ohio River Division of the U.S. Army Corps of Engineers is navigation. Figure N-14 shows a schematic elevation of the Ohio River and the location of the final configuration of 19 locks and dams that will provide year-round navigation depths on the river. The Corps of Engineers is replacing a series of 46 low-lift locks and dams with 19 high-lift locks and dams. Most of the work in the upper and middle river has been done, and present work is emphasizing the remaining work on the lower river. These structures are expensive and have been the subject of considerable value-engineering studies. A few of these studies are described in the following sections.

Uniontown and Newburgh

The Uniontown and Newburgh projects are the first replacement structures on the lower Ohio River. On the upper river projects, the locks plus the

Before

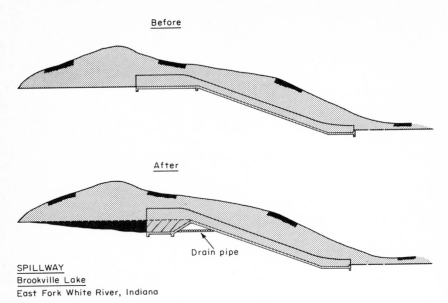

After

Drain pipe

SPILLWAY
Brookville Lake
East Fork White River, Indiana

FIGURE N-12
Profile of Brookville dam spillway area before and after value-engineering studies.

FIGURE N-13
Brookville dam during construction.

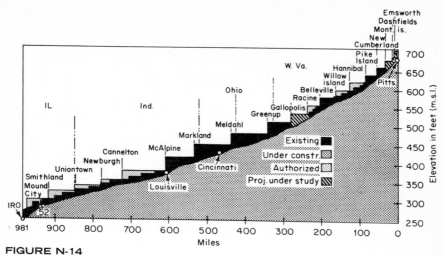

FIGURE N-14
Schematic elevation of the Ohio River and location of final configuration of locks and dams.

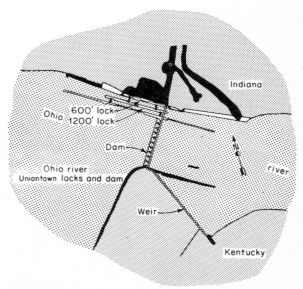

FIGURE N-15a
Uniontown project under construction.

FIGURE N-15b
Uniontown project under construction.

gated spillway and the small powerhouse usually filled the river channel, with perhaps only a short fixed-weir section.

Because of the extreme width of the lower river channel, the fixed weir became an expensive item. Figure N-15 shows Uniontown during construction. The fixed weir shown here extending from the lower left to the cofferdam is 2,239 ft long and includes the first gates under construction.

Figure N-16 shows Newburgh under construction. The fixed weir was designed to be 1,380 ft in length and extend to the gap in the river channel to the left of the cofferdam.

The weir section for Newburgh was designed with sheet-pile cells capped with concrete with a hydraulically shaped crest (Fig. N-17). The value-engineering study team first questioned the need for a hydraulically shaped crest, since the difference in water levels would be only about 1 ft when the weir was passing flow.

The value-engineering team next questioned the sunken level of the tremie concrete. Figure N-18 shows the comparison of the Newburgh design before and after the value-engineering study. The crest was reduced in size and complexity. The tremie concrete cap was raised 7 ft to make forming easier. The total savings amounted to $1,126,400. The breakdown of the savings was as follows:

Concrete eliminated	$ 515,800
Previous fill increased	18,200
Waterstop eliminated	41,900
Reinforcing steel (dowels) added	5,300
S-32 and S-28 piling reduced	37,000
Total	$ 571,200
Application of study to uniontown design:	
Concrete eliminated	$ 525,400
S-28 piling reduced	38,700
Reinforcing steel (dowels) added	5,500
Total	$ 558,600
Cost of value-engineering study, drawing, and specification changes	$ 3,400
Grand total	$1,126,400

FIGURE N-16
Newburgh project under construction.

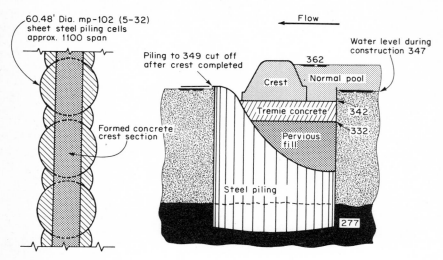

NEWBURGH LOCKS & DAMS
Ohio river

60.48' Dia. mp-102 (5-32)
sheet steel piling cells
approx. 1100 span

Piling to 349 cut off
after crest completed

Water level during
construction 347

Flow

362

Crest   Normal pool

Tremie concrete   342

332

Formed concrete
crest section

Pervious
fill

Steel piling

277

FIGURE N-17
Newburgh project—fixed-weir section.

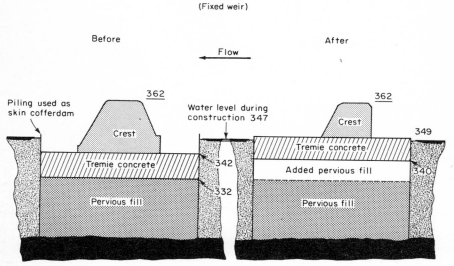

NEWBURGH LOCKS & DAM
Ohio river
(Fixed weir)

Before   After

Flow

362   362

Piling used as
skin cofferdam

Crest   Crest   349

Water level during
construction 347

Tremie concrete   342   Tremie concrete

Added pervious fill   340

332

Pervious fill   Pervious fill

FIGURE N-18a
Comparison of Newburgh design before and after value-engineering studies.

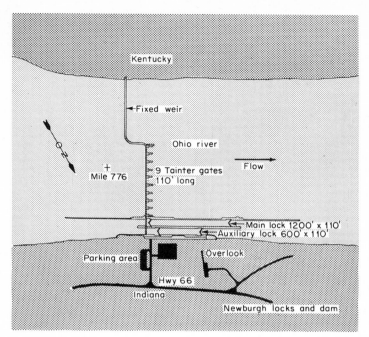

FIGURE N-18b
Newburgh under construction.

Racine Locks and Dam

The dam for the Racine locks and dam was designed by the Huntington District to be constructed with a standard three-stage cofferdam. Figure N-19 is the Racine locks and dam, completed in December 1967, in the middle part of the river. During construction, a contractor-inspired value-engineering change resulted in considerable savings to the government and extra profit for the contractor.

Under the value-engineering incentive clause of their contract, the joint venture contractor (Johnson-Kiewit-Massman) proposed a two-stage cofferdam method of construction instead of the specified three-stage. Figure N-20 illustrates the two-stage approach.

This change reduced the cost of performance of the work by $442,970. The government share of these savings was $221,485, and after the deduction of the offsetting design and review costs for government personnel, the net cost reduction resulting from this value-engineering proposal was $217,938.

As a spin-off of this VECP, the design for the Willow Island dam downstream from Racine was changed (in-house) to utilize the two-stage cofferdam method of construction. This resulted in a savings accruing to the government of $1,180,500.

FIGURE N-19
Racine locks and dam.

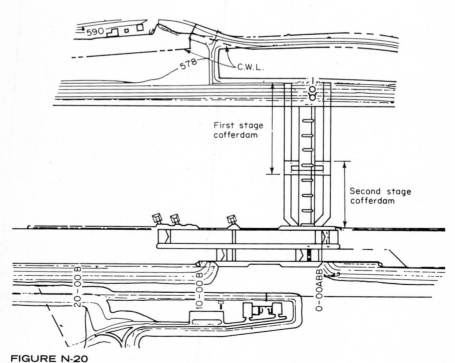

FIGURE N-20
Two-stage cofferdam method of construction.

Smithland Locks and Dam

Smithland locks and dam in the lower river is 63 river miles from the mouth at Cairo. This is the largest dam to be built on the Ohio, estimated to cost about $200 million—a prime candidate for value engineering.

The design for Smithland by the National district included 17 Tainter gates 29 ft high × 110 ft wide. These were easily the most expensive portion. Also, the Smithland dam included two 1,200-ft locks and a 900-ft fixed-weir section. The value-engineering team selected the most expensive item on the project for investigation—Tainter gates. The team conducted extensive studies to determine means to reduce the cost of this feature. These studies included taking of additional core borings and engineering calculations for several alternative layouts of the dam. The study required about a one-year effort. The value-engineering team recommended that the height of the Tainter gates be increased 29 to 36 ft, thus increasing the discharge capacity for each gate. With this increase in discharge capacity, the number of gates could be reduced from 17 to 11.

The value-engineering studies were verified with a hydraulic model test and included an increase in the length of the fixed weir, as well as a considerable increase in channel excavation, mostly in rock, both upstream and downstream of the gated section.

However, despite the increased cost of excavation and fixed weir, the design as proposed by the study team permitted construction of the entire gated structure within a single cofferdam rather than in the original two stages. This reduced the construction time and cost. Also, the founding depths for the piers eliminated would have been excessive and therefore expensive.

The net reduction in construction costs, after deduction of cost of the value-engineering study, which was $129,000, was estimated to be $19,271,000. This was the largest single value-engineering study of the Corps conducted through 1973. It has been validated and represents 110 percent of the Corps' 3-year value-engineering goal—a magnificent accomplishment by the Nashville District and the Ohio River Division value-engineering teams.

IMPLEMENTATION

After a value analysis has been completed and presented to management, the next stage is implementation. The genesis of the analysis will have a substantial impact upon the readiness to implement.

Another implementation control will be the owner's preconception of worth. In highways, for instance, the New Jersey traffic circle, with its wheel-spoke road relationship, is one of the most efficient in terms of initial cost but worst in terms of handling traffic and preventing accidents. If the Highway Department guidelines overrule traffic circles, a value analysis suggesting circles would be quickly refused, unless it were innovational.

In elementary school construction, the combination auditorium-cafeteria-gymnasium is highly cost-effective. In high schools and colleges, this combination facility is counterproductive. A value-analysis solution leading out of the well-worn tracks would have to be very persuasive, and even after the idea has been presented, it is just another study, even if an approved one.

The TFX-111 fighter-bomber taught us that an aircraft performing the functions of two separate planes may well cost more than the two planes combined. The evolution in design of stadiums has imposed very difficult design problems upon a stadium which is performing as both a football and baseball stadium. Also, in August, September, and October, the overlap between the two seasons may preclude total utilization. At the State of New Jersey Sports Complex at Hackensack Meadows, the decision was made to return to separate baseball and football facilities, with the value justification on the basis of higher quality and better revenues.

In *Building Design & Construction* (Ref. 1) there is a report of a value-engineering workshop review of Southside Medical Hospital in

Pima County, Arizona. The workshop trimmed the overall cost estimate from $32 million to $26 million by trimming square footage, interstitial space, and floor-to-floor height. Nearly $5 million of the projected $6 million in savings resulted from reducing square footage to fit program limits. Deletion of the interstitial space was estimated to save $200,000, and lessening the floor-to-floor heights resulted in a total reduction of 2 ft in the five stories at an additional saving of $100,000.

The workshop session was conducted by C. M. Associates, Sundt Construction Company of Phoenix, and the architects Friedman and Jobusch. An interesting comment was that $2 million additional could be cut if the owner "elects to streamline the project even further." This implies that the owner has some worth considerations not necessarily shared by the value-analysis team.

In the *Engineering News-Record* (Ref. 2), a report on inflation noted that at the University of Texas Special Events Center, high bids ($8.5 million over estimate) caused review of the scope and elimination of certain features such as handball courts, quality of backfill, precasting of retaining walls, and certain luxuries. This reclassification indicates that in the eyes of the owner, the features were secondary functions. An important implementation consideration would be the fact that the time lost in rebidding, and its attendant inflationary increase, could well be greater than the combined value of the reductions.

THE CONTRACTOR

Where the project is still in the design phase, the only contracting input occurs either as a courtesy or when the contractor is part of the construction-management team. In the latter case, the contractor becomes part of the owner's professional team and falls into the same constraints as the architect-engineers.

On projects where there is a VEIC (value-engineering incentive clause), the contractor may submit a VECP (value-engineering change proposal).

Submission of a VECP must allow sufficient time for implementation. The federal agencies which normally have VEICs are very cognizant of the adage "time is money." Accordingly, a valid VECP must permit sufficient implementation time.

In the 1973 GSA/PBS report on the VECP program, the four following VECPs were withdrawn because there was insufficient time allowed for implementation.

1. Exterior utilities: pumping station. The contract called for a reinforced concrete building to contain pumping station equipment. The contractor proposed to substitute a prefabricated steel underground station. The contractor later withdrew this proposal, stating that there was insufficient time to order the package station in time to meet contract completion date.

2. Mechanical-electrical equipment: elevators. Contract plans called for a standard electric elevator to be installed. Since the structure was only two stories plus basement, the contractor proposed installation of a hydraulic elevator. This proposal was questioned because of a high water table which could provide leakage problems at the cylinder. The contractor withdrew the proposal as details could not be worked out in the short time allowed by this procurement schedule.

3. Caissons. The contractor recommended elimination of heavy permanent casings and the use in place of these of full-length temporary personnel protection liners. The initial response was that the VECP would not be acceptable because of soil conditions. However, if after conferences with the structural engineer it would be possible to eliminate some of the liners, VECP would be reconsidered. "There was not enough time for a review and the VECP was withdrawn."

4. Redesign of HVAC systems. The contractor submitted preliminary design intent; "however, since complete redesign would substantially delay the contract completion, VECP was withdrawn."

The following comments are from the Public Building Service in regard to submission of VECPs (Ref. 3):

VE is not easy. It is a challenging way to reduce life cycle costs or the cost of your contract.

To have a valid VECP, you must do the following things in the proposal: identify it as a VECP; cite the VE Incentive Clause in your contract; recommend a specific change to the contract; offer savings to your contractor; show where PBS will save money over the life cycle.

VE does not come for free. PBS is asking that you invest your resources in developing and submitting VECPs. You will lose this investment each time a VECP is disapproved. In addition, you cannot delay performance under your contract while a VECP is in process.

Divulgence of your idea will not jeopardize your right to a fair share of your savings. When you do discuss your idea with a PBS representative, we suggest that you follow these discussions with a letter of intent to submit a VECP.

Format . . . no special forms are required. You may use your company letterhead and estimating forms. Don't forget to identify your proposal as a VECP.

In the case of PBS VECPs, normal procedure is a letter of direction to proceed subject to cost and time limitations. This is a legal notification of intent to alter the contract and the contractor's authorization to proceed. It describes the parameters of the negotiations for cost and time. These will result in a formal contract change-order document.

Thus, implementation of the VECP evolves into a standard contractual change-order situation.

In a non-VEIC clause contract, the contractor has much less motivation to submit a VECP. Acceptance of a VECP will result in a change order, but the change order would presumably be a net credit to the owner. In states and governmental units where value analysis and value engineering have not been recognized by legislature, the government contracting officers have no authority to divide savings between the contractor and the owner. All savings, thus, accrue to the owner, to the actual detriment of the contractor. The contractor, after all, entered into a contract for a fixed amount. A reduction in the contract should be limited to the change itself and should not attempt to take away general conditions or overhead monies which the contractor had allotted to the job.

Obviously, in most cases, the contractor has no real incentive to submit this type of a VECP. The result is a difficult negotiation for a credit change. (Many contracts contain add-deduct units, with a lower cost for a deducted unit and a higher cost for added units of work—thus reflecting the reality of the contractor's loss in handling credit change orders.) There are, of course, occasions where a contractor recognizes a particularly attractive value improvement wherein perhaps actual costs are reduced while delivering equal or greater value. In this situation, the contractor may request a substitution at no change in cost. If the value provided is equal or greater, the contracting officer can accept this, recognizing that the contractor is also benefiting. There are also occasions where a contractor in a non-VEIC environment can submit an increase in cost based upon better life-cycle costing factors. The gain might be in the availability or ease of installation of better equipment.

THE DESIGNER

The greatest opportunity for value analysis is in the preconstruction phase. Much has been discussed in regard to the possible addition of increased incentives for the professional architect-engineer to conduct value analysis. However, the continuing requirement which professionals impose upon themselves is to design for optimum value.

Accordingly, value analysis becomes a normal part of the design cycle, or should. When the owner requires the designer to perform value analysis in strict accordance with the disciplines, in particular functional analysis in a creative environment, the owner may choose to provide additional compensation. In *Civil Engineering* (Ref. 4), Chris Law, director of professional services, GSA, is quoted as saying that "compensation for value engineering service is separately identified and may be excluded from determining compliance with the statutory 6% limit on professional fees."

Reference 5 reports the Joint Committee expression of opinion on the role of architect/engineers in administering value engineering. This committee is part of the National Public Advisory Panel on Architectural Services and is made up of nine professional members, with a predominance of representation from the American Institute of Architects (AIA):

> AEs can play a significant role in assisting the Government to achieve better value in their construction and, therefore, will be responsive to their new opportunities in value engineering. However, we will require well defined procedures to effectively implement our role in the value engineering process. The application of value engineering by the AE should proceed systematically with the design. We see value engineering assisting to provide better cost control and allocation of funds within the design.

> We are strongly against receiving incentive shares resulting from value engineering savings. It is our obligation to provide the most economical design. However, present fees do not provide the resources to do everything necessary in the design process to achieve these economies. Additional fees for value engineering effort would be beneficial to both parties and must be provided.

> If value engineering becomes a specific requirement in the design contract, AEs will seek and achieve qualification in this area. We encourage AE participation in Government value engineering training courses and consider it reasonable for education of AE personnel to be required as part of the AE scope to perform value engineering.

> In addition, AEs should be allowed to review contractor proposals to maintain integrity of design and insure that life cycle costs are considered rather than just initial costs.

The Public Building Service has several levels of value-engineering services:

1. Basic value-engineering services for architect/engineer contracts

2. Moderate value-engineering services for A/E contracts

Table 14-1
Moderate Value-Engineering Services for A/E Contracts

A. DESIGN CONCEPT STAGE

102. At an early time, the A/E shall host a one day Executive Seminar in VE using a qualified VE Consultant.

    a. The Seminar shall be attended by a suitable number of employees from the A/E's firm, and from each design consultant firm. In addition, the A/E shall invite the attendance of representatives from the Government and the various using agencies of the facility under design.

    b. The Seminar schedule and program of instruction shall be submitted to the Contracting Officer for approval at least two weeks in advance of the proposed commencement of the Seminar.

    c. The A/E shall provide suitable seminar facilities and program for participants in the Seminar session.

    d. The Government will provide guest speakers as desired and certificates for each participant.

103. The A/E shall examine all design criteria for each of the various disciplines of work; i.e., site, architectural, structural, mechanical, and electrical, furnished under the contract, for the purpose of identifying and questioning constraints to achieving the required task at the lowest overall cost consistent with desired performance requirements.

    a. Based upon this examination, the A/E shall submit a report to the Contracting Officer identifying areas where criteria changes are considered desirable to develop maximum savings in structures, equipment, materials, or methods, even though such recommendations may be at variance with existing GSA criteria or instructions provided concerning the design in question.

    b. Recommended criteria modifications, together with the magnitude of savings therefor, should be included in the report. Use of GSA Form 2762, Design Review Ideas, is suggested.

    c. This report shall be submitted with the design concepts.

B. DESIGN TENTATIVE STAGE

111. The A/E shall perform a task team effort to review his tentative design submittal for VE ideas and cost effectiveness. Immediately upon completion of the task team effort, the A/E shall submit a report of the teams's VE ideas to the Contracting Officer. Use of GSA Form 2762, Design Review Ideas, is suggested.

C. INTERMEDIATE WORKING DRAWING STATE

121. The A/E shall perform a task team effort to review his intermediate working drawing design for VE ideas and cost effectiveness. This review should concentrate on high volume use items provided in the design such as doors, valves, finishes, etc. where the cost to make a change is minimal compared to the potential savings.

D. POST CONSTRUCTION CONTRACT AWARD STAGE

130. The Government will receive and initially screen VECPs submitted by the construction contractor(s) pursuant to the VE Incentive Clause in his contract.

    a. The A/E will be asked to review and comment on those VECPs the Government is considering favorably.

    b. If the A/E considers the adoption of a specific VECP requires design effort beyond the scope of this contract, the A/E shall so notify the Contracting Officer in accordance with Clause 3 of the General Provisions of this contract. The Contracting Officer will then determine whether or not the additional service will be required.

F. STANDARD SERVICES

140. The conducting of VE Workships as required by this contract shall be performed by an outside VE Consultant. The A/E shall submit for approval of the Contracting Officer the name(s) of the VE Consultant (with their record of education and experience) whom he proposes to provide the workshop leadership required by this contract. The VE Consultant shall be qualified by education, training, and experience equivalent to that required by the Society American Value Engineers for certified value specialists and shall have a recognized background in the field of Value Engineering.

142. The VE task team required by this contract shall be composed of at least one architect, structural engineer, mechanical engineer, and electrical engineer with other support as necessary. Individuals on this team should not be the same persons responsible for the original design work. The A/E shall submit for approval of the Contracting Officer the names of the members of the VE task team (with their record of education and experience) whom he proposes to perform the VE design reviews required by this contract. The A/E can meet this requirement by one of the following three options:

    a. Utilize employees of the A/E and his design consultants, all of whom have completed a 40-hour VE workshop course accredited by the Society of American Value Engineers or its equivalent.

    b. Engaging a qualified VE Consultant to work with and guide the inexperienced employees of the A/E's firm and his design consultants. In this option the VE Consultant would serve as team leader.

    c. Engaging a qualified VE Consultant to provide the full service. The VE Consultant shall be qualified by education, training, and experience equivalent to that required by the Society of American Value Engineers for certified value specialists and shall have a recognized background in their field of Value Engineering.

144. The A/E shall review and consider for incorporation VE changes and ideas suggested by the Government, by his VE review team, or by his VE Consultant. The A/E shall take one of the following courses of action for each idea suggested:

    a. During design, incorporate the idea provided that it is convenient to do so at no additional fee and so advise the Government.

    b. Reject the idea and advise the Government as to the reason.

    c. Recommend approval of a change to completed and approved design work or GSA criteria and advise the Contracting Officer of the additional fee necessary to incorporate the idea in accordance with Clause 3 of the General Provisions of this contract. The Contracting Officer will then determine whether or not the additional service will be required.

    d. Use of GSA Form 2777, Summary Report, is suggested to report disposition of VE ideas.

147. The Government will furnish VE Workbooks, Executive Briefs, and all other referenced forms for the information and use of the A/E as necessary.

G. HANDBOOKS

150. VE Handbook, PBS P 8000.1, will be issued for the information and use of the A/E and his design consultants.

H. FEE AND PAYMENT

160. With regard to VE services, no incentive payments to or sharing of savings with the A/E will be made by the Government in connection with this contract.

SOURCE: Public Building Service/GSA Manual P 8000.1, *Value Engineering.*

3. Extensive value-engineering services for A/E contracts

4. Special value-engineering services for A/E contracts (for which a construction manager is assigned)

5. Value-engineering services for construction-manager contracts

Table 14-1 is the description of moderate value-engineering services for A/E contracts as listed in Ref. 5.

In Design for Value (Ref. 6), the Public Building Service answers some frequently asked questions:

> Does the A/E or CM receive an incentive share of any savings that might result? No, we consider VE another professional service which is fully compensated by the A/E or CM contract fixed price.
>
> Won't reducing the cost of the project also reduce the amount of A/E's fee? . . . No, the A/E's fee is a fixed fee negotiated at the start of the contract.
>
> Isn't VE second-guessing? . . . No, taking another look from a different point of view is good business and is healthy for any designer.
>
> Won't VE increase the design time? . . . It shouldn't, if you make your value reviews concurrently with your normal design reviews. In fact, VE often eliminates or simplifies design work before it is done. It often prevents the delay that occurs after bids are received and the project is found over the budget.

There are clients who by choice or by ordinance or legislation tie the design fee to the actual cost or the estimated cost, whichever is smaller. In this situation, the designer is reducing his own fee when he submits valid cost reductions through value analysis. The best answer to this is to prenegotiate a design fee, as described in the PBS approach above.

THE SCHEDULE

The schedule of implementation is vital. The successful VECP or design proposed change, if recognized early enough, will not cause delay. It will, then, have no scheduling implications. However, the quality of schedules varies widely in different projects, and the time-cost relationship is so vital that the value team should insist upon the availability of a viable schedule.

Figure 14-1 illustrates a breakeven point for a proposed 4 percent savings. If the project is delayed 4 months, this savings will break

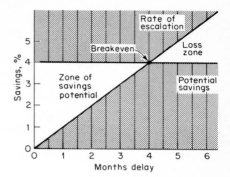

FIGURE 14-1
Breakeven chart. (Public Building Service/GSA.)

even at best, and this does not include the cost of preparing the proposal.

In addition to escalation, the cost of delay should also include loss of revenue and/or utilization.

Each project, either through designation or performance, has a project manager. It may be the architect, the owner, or a designated project manager such as the construction manager. This project manager should have charge of the project schedule and its implementation. Where definitive techniques such as the critical path method (CPM), PERT, or precedence diagramming are utilized, the value team can make a definitive evaluation in regard to the timing of the value change. Time for redesign and approvals should be incorporated into the time allowed for implementation of the change.

Experience indicates that the design period will essentially equal the construction period. If administrative reviews are timely and concurrent, they will not add preconstruction time. However, particularly in governmental situations, review times may be mandated and may require the interruption of the design before it may proceed. Usually, the designers can proceed at their own risk and without progress payments. The tailoring of value-analysis of the concepts and scope of each of the design phases has been previously covered and underscores the need to start value analysis at the earliest possible point in the design process. At the twelfth SAVE meeting, Max O. Urbahn, past president of the AIA, suggested that value analysis should start "as soon as the design process begins." G. D. Fremouw, director of HEW's Facilities Engineering & Construction Agency (FECA) said, "the sooner, the better."

During construction, scheduling is usually the responsibility of the contractors—and the concern of the owner and the project manager. Where a single construction contractor is handling all of the work, that contractor is key to all scheduling problems and solutions. In some forms of construction management, the construction manager takes over the role of the general contractor and subcontracts the work to many contractors.

Contractors schedule in a broad range of ways and degrees. The successful ones utilize the best techniques and manage aggressively. The worst ones do not stay in business long, but while they do they cause problems. The majority of contractors are somewhere in between.

To the sophisticated contractor, a change resulting in a shorter contract duration should result in a VECP with mutual savings, even if the cash value of the contract labor and materials is not reduced, because his percentage of overhead should be reduced.

SUMMARY

Value is often a relative thing and varies with the beholder. In one design situation involving a series of academic buildings valued at about $50 million, it was mutually agreed that value analysis would be used during the design phase. The construction manager, with overall responsibility for controlling scope and cost during design, suggested value-analysis reviews as one of the means of controlling costs.

The individual college saw value analysis as the opportunity to get more building for the budgeted amount of dollars; it had no intention of returning dollars unused. The state-level construction agency saw value as building the same for less dollars so that they could return some monies to their organization unexpended. The design teams saw value as providing the basic program for lower dollars, leaving some flexible amounts available for increasing aesthetic treatments, without a risk of overrun on the basic scope. (Overrun on the basic scope would involve the risk of requiring redesign during an escalating cost period—a vicious cycle.)

REFERENCES

1. *Building Design & Construction,* Aug. 1974.

2. *Engineering News-Record,* July 25, 1974, p. 10.

3. *Build Your Profits,* GSA Pamphlet, DC 74-6961.

4. *Civil Engineering* (ASCE magazine), Nov. 1973, p. 14.

5. Public Building Service/GSA Manual P 8000.1, *Value Engineering,* U.S. Government Printing Office, Washington, D.C., Fig. 7-21.2, p. 7-7.

6. *Design for Value,* GSA Pamphlet, DC 74-587.

VALUE ANALYSIS IN
THE CONSTRUCTION CONTEXT

THE OUTSIDER

Over the years, designers and constructors have evolved into adversarial roles in carrying out their individual responsibilities during the life of a construction project. However, as many police officers who have attempted to settle domestic quarrels can sadly attest, adversaries with a basic mutual regard quickly band together to face an outside third party.

Value analysis can, and should be, applied to the construction project as part of a team effort. A secondary function of the value-analysis sessions should be the development of a better communication within the team. Unfortunately, value analysis imposed upon the parties to the construction project without suitable understanding of their roles, responsibilities, and longstanding practice of thorough analysis can be a counterproductive factor.

In testimony before a congressional committee, Larry Spiller, the American Consulting Engineer Council's assistant executive director stated (Ref. 1): "The selection of topnotch designers, plus allowance of an adequate fee, will have as much, or greater, impact as performance of value analysis at the conclusion of design."

The statement is unfortunate since it seems to imply that value analysis is in contention with, or upsetting to, the selection of excellence in the design team at appropriate fees.

Value analysts tend to draw negative reactions and defensive response when making speculative comments on the inefficiencies of the construction industry as a whole. For instance, one value specialist remarked that it was obvious that schools waste money on poorly designed heating systems. As an example, he noted that in the spring and fall, school windows on many schools in the area are open because the classrooms are too warm. Obviously, he comments, this is a poorly designed system and one which is wasting heat.

On the surface, it would appear that he is right. However, it is also very possible that he is *absolutely wrong*. Many schools are budgeted so tightly that only a very basic heating system is installed. In more recent schools, this is usually perimeter fin tube heating at the window line. Heat flowing through the loop to remove the chill from the last room on the heating loop flows through other classrooms. When the heating demand is very low, this heat flowing through generates more heat than is desired in the individual rooms.

Certainly, the situation is undesirable. However, if under analysis it is determined that the situation exists for only a very limited number of days, the life-cycle cost of the initial installation and the fuel utilization could well be efficient from a cost standpoint. Value analysis probably indicates that a modest investment in more heating zones, insulation, and convectors permitting more local control would be worthwhile. It is probable that the analysis would not disclose enormous savings in either energy or initial cost in this particular area. In a case history, The Electric Energy Association (EEA) posed an even more startling situation which would appear to be poor value. A conservation-minded citizen passing a school building late on a dark windy night in midwinter is startled to see the building's lights flash on suddenly. This suggestion of some type of misdesign and energy wastage would in fact have been the routine cycling of the lighting system to maintain temperature in a heat-recovery system commonly used in building design. In this type of system, waste heat from the lights is recovered and used to maintain the building temperature. The EEA case history notes that a typical reply

> . . . would explain the energy conservation features of the HVAC system stressing, for example, heat reclaimed, economizer controls, program ventilation cutoff, and automatic temperature setback. Assurances would be given that cycling the lights is an economical means for preventing indoor temperatures from dropping below a practical minimum and is perfectly consistent with the energy management objectives of the designers.

That would be the typical answer, but the designers would probably not have a convenient or timely forum in which to offer the answer. Accordingly, the designers in the Piscataway High School in New Jersey, the subject of the case study, used a different approach for their off-hour heating requirements. They utilized (1) the thermal inertia of the water in the hydronic network, and when that diminished, (2) the heat stored in the off-peak hot water generator, and finally, (3) the supplementary strip heaters in the air handling units. This system, which is also energy-conserving, avoids the poor public impact of lights flashing during the night.

While the Piscataway solution is certainly a better one, the value analyst is well advised to withhold comment until sufficient information is available for suitable consideration of a situation.

One of the important techniques utilized by value analysts is creativity, and the importance of maintaining an open mind is continually emphasized. In the spirit of "physician heal thyself," the value analyst is well advised to approach the construction-design-build team with an open mind, and with respect for their past efforts and competence—until the facts prove to the contrary.

In Ref. 2, a snap judgment might have indicated that the A/E firm of Reid and Taris Associates was being extravagant when they specified marble floors at the Civic Center subway station of the BART Transit System in San Francisco. The firm also called for an alternative bid for terrazzo. Contrary to usual experience, the marble floors, which provided the aesthetics and value really desired, were bid $110,000 less expensive than terrazzo. The saving was a result of an unpredictable interindustry competition.

WHAT'S IN A NAME?

As discussed previously, the name value engineering evolved from value analysis principally because of its use in the Navy Department and the need for categorization of the type of professional work involved. In that context, the evolution was appropriate and useful.

Unfortunately, the name in the construction industry tends to be limiting. While it could be argued that the term engineering and the practice of engineering are integral to project design and construction, the definite implication is that value analysis falls within the province of the engineer. This would have as a correlary that it is not in the province of the nonengineer. This position is taken very seriously by leading members of the industry. William A. Carlisle, AIA, a member of the Government Affairs Commission of the American Institute of Architects, made the following statement regarding the term to the Subcommittee on Public Buildings and Grounds of the Senate Committee on Public Works:

> Today [June 19, 1973], the American Institute of Architects, the National Society for the architectural profession, representing 24,000 licensed architects, wishes to express its views on the use in federal construction, of value engineering. For the purpose of these hearings, we will use the

term—value analysis—which we believe is more appropriate and more descriptive of the process presently under this Committee's review.

Many others, including Representative Larry Winn, Jr., prefer to call value analysis value management. The Public Building Service of GSA has changed the title of its value-analysis group from value engineering to *value management*.

In Ref. 3, the Department of Defense handbook states that value engineering (VE) is the term used in this handbook and by the DOD in its contracts. "Others may refer to their value improvement efforts by such terms as value analysis, value control, or value management. There may be some subtle differences between these other programs and VE, but the basic objectives and philosophy appear to be the same for all."

The choice of the proper term to describe value analysis in the construction industry is important from the positive viewpoint also. A prominent member of the AIA commented: "Value engineering is just part of basic project cost control." And an officer in the American Association of Cost Engineers states: "Obviously value engineering falls under what we know as cost engineering." While these statements are not necessarily incorrect, they lead to a misconception discussed in Ref. 4: "Value engineering is nothing more than good design engineering."

The DOD book goes on to say that this is an unjustified simplification. "VE draws on each of the proven disciplines, in different circumstances, . . . all set forth as a formal, orderly investigative procedure." While the quantum jump to a solution should not be precluded or ignored, value analysis *must* include a disciplined job plan, and this job plan must be built around functional analysis with development of alternatives in a creative atmosphere.

In other applications, value analysts have taken great care to describe with precision the relationship of value analysis to existing programs and techniques. For instance, in Ref. 5 the Department of Defense correlates value analysis with program (project) offices; cost effectiveness; systems analysis; configuration management; standardization; zero defects; reliability, quality assurance, and maintainability; life-cycle costing; and integrated logistic support.

In Ref. 6, the Public Building Service discusses the relationship of value engineering to other programs and disciplines. This review covers management-improvement programs; cost reduction; cost effectiveness; systems analysis; standardization; zero defects; reliability, quality assurance, and maintainability; life-cycle costing; and tradeoff analysis. Essentially, the PBS evaluation has taken the DOD analysis and extended it into the construction industry.

In *Value Management* (Ref. 7), Heller explains the need for the establishment of the context of programs in the design phase: "designers are complaining . . . today . . . that they are getting so much help from . . . specialists that they cannot get their jobs done. From the designer's viewpoint, ideally he should be allowed to do his own job, recognizing that under ordinary circumstances his capabilities in these many fields will suffice. In those exceptional circumstances in which the designer faces a problem beyond his capability, he should have available to him (but not forced on him) the skills of the various specialists upon whom he can call at will. The only trouble with this arrangement, as many specialists will be quick to point out, is that sometimes the designer does not recognize when he is in need of help."

QUALIFICATIONS AND TRAINING

The role of value analysis in the construction industry is evolving. Presently, the training of value specialists has been sponsored principally by the Society of American Value Engineers (SAVE) and the Public Building Service.

SAVE has established a program for testing and certifying value specialists. A written examination is given for qualification as a certified value specialist (CVS) by SAVE.

Commissioner L. F. Roush of Public Building Service (Ref. 8) indicates that GSA requires the use of certified value specialists or equivalent in the instruction of value-management workshops and executive seminars. These workshops and seminars are required under the GSA design contracts.

For designs over $3 million, GSA requires that a CVS or equivalent lead the value task teams, unless all members of the task team have been to an accredited value-analysis workshop.

The PBS office of value management maintains a file on the qualifications of consulting value specialists who are interested in working with PBS A/E firms in training and/or value-analysis reviews.

Presently, there are only a limited number of value analysts experienced in the application of value analysis in the construction industry. While SAVE has actively encouraged and supported the development of value analysis in the construction industry, the majority of the SAVE membership has not been involved in day-to-day activities involving design or construction. Accordingly, many of the value specialists working with task force teams must become educated in the procedures, practices, and requirements of the typical construction project at the same time as they are providing their expertise in the discipline of value analysis. This is a growing period in the evolutionary cycle and will result in the development of value specialists into knowledgeable construction value specialists, and similarly, many designers and constructors will acquire the techniques and disciplines of value analysis and become construction value specialists in their own right.

This evolutionary process has been experienced in the construction industry many times before. One of the most analogous situations was the development and evolution of computerized planning for cost and scheduling known as the critical path method (CPM). This method was developed specifically for the purpose of applying computerized planning in the industry. The relative success of CPM (and PERT) applications demonstrates that the industry is willing to accept those techniques which have a common-sense basis and produce results. The role of the specialist in encouraging and maintaining the evolution has been indispensable. However, it should be noted that on an individual basis, many people in the industry have been disappointed and even embittered by encounters with methods analysts, operation research specialists, and a variety of consultants who seek to improve what appear to be inefficient methods of design or construction but fail to recognize the many constraints imposed upon the delivery of construction projects by existing rules and laws. Those experts who chose to learn about the construction industry and work within the practical constraints have produced many improvements, but those who persisted in selling only their own expertise, purported objectivity, and methodology are long gone.

In *Value Management* (Ref. 9), Heller warns of this:

> Some practitioners of value engineering are falling into the trap that has caught others, such as those impressed by promises of PERT techniques and computer applications, the trap of insisting on rigid adherence to the methodology, instead of maintaining a results-oriented frame of mind . . . We have become obsessed with, or perhaps just been overwhelmed by, techniques and methodology . . . . We have a tendency to misuse them. A great problem with these techniques is the tendency for overuse.

THE INTERNAL GROUP

The owner, designer, and contractor should each consider the development of an internal value-analysis staff. Depending on the size of the organization, this could be either an ad hoc group or a permanent staff group.

At the Public Building Service, there is a permanent value-management staff supplemented by a large number of key PBS professionals who have value management as a collateral duty. The ad hoc group performs specific value-management tasks, including the constitution of value-management review boards in the various PBS regions for the consideration of VECP proposals from contractors.

In fiscal year 1973, PBS employees initiated 87 value studies. Implementation of the internal program for that year cost $318,600 and the savings were $666,500, or a return of $2 for every $1 invested. However, the true savings could well extend far beyond. For instance, the V-E summary below shows a value recommendation which produced a savings of $6,100 as a result of a $600 study. But the more important consideration is the potential application of this same idea to many other PBS designs. The internal value-analysis group can be a catalyst in the development of value criteria and suggestions. This, in turn, assists the AE in the implementation of value analysis during design.

The following summary, taken from a V-E report prepared in August 1972 on construction of a Post Office building in Midland, Texas, is a good example of savings available from substitution of less costly components that do not sacrifice required performance.

1. The specifications require 7,500 AMP interrupting capacity circuit breakers for the 120/208 volt panels. The drawings require only 5,000 I.C. breakers. The panel board change is as follows from Type NAB to NLAB panel and from Type TE circuit breakers to Type TOB circuit breakers. This change in panels and circuit breakers will in no way lower the protection required for the branch circuits. It should be pointed out that there are cheap 5,000 AMP I.C. breakers on the market. These that are proposed by the contractor are acceptable.

   Gross earnings: $1,000

2. The project specifications require a NEMA Class III switchboard with power circuit breakers. Catalogue literature shows that both Class II and Class III have the same ratings, load capacity, interrupting capacity, etc. The only difference between the two classes is that Class II boards have individually mounted devices (circuit breakers) and Class II has group mounted devices. The NEMA Class II switchboard proposed under the VE proposal has fuse switches by General Electric Type QMR. These fuse switches will meet or exceed the protection requirements of the specifications. The only disadvantage of fuse switches is the replacement of blown fuses. This is not a problem as the fuses only provide feeder protection. The connected loads have their own protective device so it is not likely the building will have a problem. The fuse type switch is a simple device and does not have to be recalibrated as the power circuit breaker does. In short, there is no advantage to have the NEMA III boards.

   Gross savings: $5,100
   Total savings: $6,100

Implementation: A change order issued to Area Builders to receive credit has been approved. The employee's VE suggestion is hereby cross fed to PBS VE Manager, Central Office, for all regions to have the opportunity to use this suggestion.

Costs. V-E study: $200
Implementation: $400
Total: $600

SUMMARY

In Ref. 8, Commissioner Roush looks to the future with a commentary on the present: "Our country would be better off if more people could design for value as design to cost. Unfortunately, many can do neither."

REFERENCES

1. *Engineering News-Record,* May 7, 1974.

2. *Civil Engineering,* Nov. 1973.

3. *Value Engineering,* U.S. Department of Defense Handbook 5010.8-M, Sept. 12, 1968, p. 1.

4. "Principles and Applications of Value Engineering," U.S. Department of Defense Joint Course, U.S. Government Printing Office, Washington, D.C., pp. 1–9.

6. Public Building Service (GSA) Handbook P 8000.1, *Value Engineering,* rev. Jan. 12, 1972, U.S. Government Printing Office, Washington, D.C., pp. 15–18.

7. Heller, E. D., *Value Management: Value Engineering and Cost Reduction,* Addison-Wesley, Reading, Mass., 1971, p. 87.

8. *Building Design & Construction,* July 1974, pp. 9–10.

9. Heller, op. cit., pp. 6–7.

INDEX